Nitrification and Denitrification in the Activated Sludge Process

Nitrification and Denitrification in the Activated Sludge Process

Michael H. Gerardi

WILEY-INTERSCIENCE

A John Wiley & Sons, Inc., Publication

For ordering and customer service, call 1-800-CALL-WILEY.

Library of Congress Cataloging-in-Publication Data:

Gerardi, Michael H.
   Wastewater microbiology : nitrification/denitrification in the activated sludge
process / Michael H. Gerardi.
      p.   cm.
   Includes bibliographical references.
   ISBN   978-0-471-06508-1
   1. Sewage—Purification—Nitrogen removal.   2. Nitrification.   3. Sewage—
Purification—Activated sludge process.   I. Title.
TD758.3.N58 G47 2002   2001046765

To
L. Vernon Frye
and
the men and women of the
Williamsport Sanitary Authority
and
Williamsport Municipal Water Authority

The author extends his sincere appreciation to
joVanna Gerardi for computer support
and
Cristopher Noviello for artwork used in this text.

Contents

Preface

Within the last 15 years much interest and use of microbiological principles of wastewater treatment have been successfully applied to the activated sludge process. These principles include the use of the microscope for process control and a better understanding of the microorganisms, especially the bacteria that are involved in the degradation of wastes.

Of special interest to wastewater treatment plant operators are the bacteria that degrade nitrogenous wastes—the nitrifying bacteria—and the bacteria that degrade carbonaceous wastes—the cBOD-removing bacteria. Both groups of bacteria need to be routinely monitored and operational conditions favorably adjusted to ensure desired nitrification. However, operational conditions do change, often in a very short period of time, and an undesired change in operational conditions can adversely affect the bacteria within the activated sludge process and its ability to degrade wastes.

Regardless of discharge permit limitations, activated sludge processes that are and are not required to nitrify and denitrify do nitrify and denitrify. Often these plants develop a form of incomplete nitrification or undesired denitrification that is responsible for an operational upset, an increase in operational costs, and noncompliance with a discharge limitation. Therefore, with a minimum of technical jargon and numerous tables and illustrations, this book addresses the microbiological principles of the bacteria and operational conditions that affect nitrification and denitrification in the activated sludge process. The book is target for operators who are responsible for the daily operation of the activated sludge process regardless if the process is or is not required to nitrify or denitrify. Each chapter is prepared to offer a better understanding of the importance of nitrification and denitrification and the bacteria involved in nitrification and denitrification. The book provides the operator with process control and troubleshooting measures that help to maintain permit compliance and cost-effective operation.

Nitrification and Denitrification in the Activated Sludge Process is the first book in the Wastewater Microbiology series by John Wiley & Sons. This series is designed for operators and provides a microbiological review of the organisms involved in wastewater treatment, their beneficial and detrimental roles, and the biological techniques available for operators to monitor and regulate the activities of these organisms.

Michael H. Gerardi
Linden, Pennsylvania

Part I

Overview

1

Nitrogen: Environmental and Wastewater Concerns

The presence of nitrogenous or nitrogen-containing wastes in the final effluent of an activated sludge process can adversely impact or pollute the quality of the receiving water. Principle nitrogenous wastes that pollute the receiving water are ammonium ions (NH_4^+), nitrite ions (NO_2^-), and nitrate ions (NO_3^-). Ions are chemical compounds that possess a negative (−) or positive (+) charge. Significant pollution concerns related to the presence of nitrogenous wastes include dissolved oxygen (O_2) depletion, toxicity, eutrophication, and methemoglobinemia (Table 1.1).

To reduce the adverse impacts of nitrogenous wastes upon the receiving water, an activated sludge process may be required by state and federal regulatory agencies to lower the quantity of nitrogenous wastes in its final effluent. The activated sludge process would have to nitrify and denitrify the nitrogenous wastes. A nitrification requirement usually is issued as an ammonia (NH_3) discharge limit, and a denitrification requirement usually is issued as total nitrogen or total kjeldahl nitrogen (TKN) discharge limit (Table 1.2).

DISSOLVED OXYGEN DEPLETION

The discharge of nitrogenous wastes to the receiving water results in dissolved oxygen depletion. The depletion occurs through the consumption of dissolved oxygen by microbial activity.

TABLE 1.1 Pollution Concerns Related to Excess NH_4^+, NO_2^-, and NO_3^-

Nitrogenous Ion	Pollution Concerns
NH_4^+	Overabundant growth of aquatic plants
	Dissolved oxygen depletion
	Toxicity as NH_3
NO_2^-	Overabundant growth of aquatic plants
	Dissolved oxygen depletion
	Toxicity
NO_3^-	Overabundant growth of aquatic plants
	Dissolved oxygen depletion
	Toxicity
	Methemoglobinemia

First, ammonium ions are oxidized to nitrite ions, and nitrite ions are oxidized to nitrate ions within the receiving water (Figure 1.1). The oxidation of each ion occurs as dissolved oxygen is removed from the receiving water by bacteria and added to ammonium ions and nitrite ions. Second, ammonium ions, nitrite ions, and nitrate ions serve as a nitrogen nutrient for the growth of aquatic plants, especially algae. When these plants die, dissolved oxygen is removed from the receiving water by bacteria to decompose the dead plants (Figure 1.2).

TOXICITY

All three nitrogenous ions can be toxicity to aquatic life, especially fish. Ammonium ions and nitrite ions are extremely toxic, and nitrite ions are the most toxic of the three nitrogenous ions.

TABLE 1.2 Permit Requirements for Nitrification and Denitrification

Requirement	Description	Nitrification/Denitrification
NH_3	Ammonia	Nitrification
NH_4^+	Ammonium ion	Nitrification
nBOD	Nitrogenous biochemical oxygen demand	Nitrification/denitrification
NOD	Nitrogenous oxygen demand	Nitrification/denitrification
TKN	Total kjeldahl nitrogen	Nitrification/denitrification

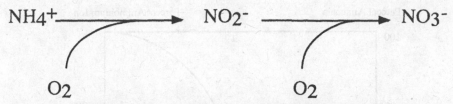

Figure 1.1 *Oxidation of ammonium ions and oxidation of nitrite ions. Under appropriate conditions nitrification occurs when oxygen is removed from water, or a water film, and added to ammonium ions to produce nitrite ions, or added to nitrite ions to produce nitrate ions. Although many organisms such as algae, bacteria, fungi, and protozoa are capable of nitrifying ammonium ions and nitrite ions, a specialized group of nitrifying bacteria is primarily responsible for nitrification in water and soil.*

Although ammonium ions are the preferred nitrogen nutrient for most organisms, ammonium ions are converted to ammonia with increasing pH (Figure 1.3). It is the ammonia at an elevated pH that is toxic to aquatic life.

EUTROPHICATION

While phosphates (PO_4^{2-}) are the primary source of eutrophication, nitrogenous wastes contribute significantly to this water pollution problem. Eutrophication refers to the discharge of plant nutrients, primarily phosphorus and nitrogen, in undesired quantities to bodies of freshwater, such as lakes and ponds. The presence of undesired quantities of plant nutrients stimulates the rapid growth or blooms of

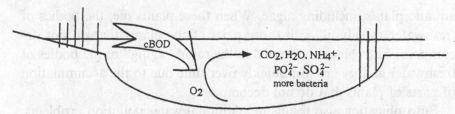

Figure 1.2 *Oxygen used during decomposition of dead plants. As large blooms of aquatic plants die in the water, a large diversity of bacteria and fungi quickly remove large quantities of dissolved oxygen and decompose the plant tissue into carbon dioxide, water, ammonium ions, phosphate ions, and sulfate ions. The bacteria and fungi transform some of the organic material from the plant tissue into new bacterial and fungal cells.*

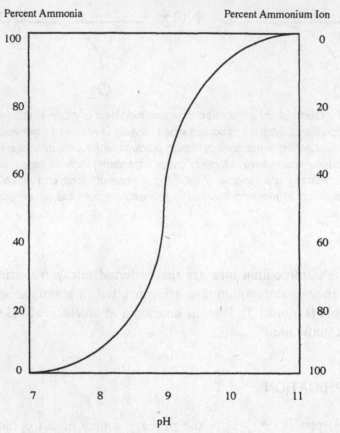

Percent Ammonia — *Percent Ammonium Ion*

Figure 1.3 *pH and the conversion of ammonia and ammonium ions. The relative quantities of ammonia and ammonium ions in water are determined by the pH of the water. As the pH of the water decreases, ammonium ions are favored. As the pH of the water increases, ammonia is favored. At a pH value of 9.4 or higher, ammonia is strongly favored.*

aquatic plants, including algae. When these plants die, the bodies of freshwater rapidly fill with those parts of the plants that do not decompose. Eutrophication results in the rapid "aging" of the bodies of freshwater as they are lost quickly over time due to the accumulation of parts of plants that do not decompose.

Eutrophication also results in additional water pollution problems. These problems include fluctuations in dissolved oxygen concentration with the growth and death of aquatic plants, the clogging of receiving water caused by the sudden bloom of aquatic plants, and the production of color, odor, taste, and turbidity problems associated with the growth and death of aquatic plants.

METHEMOGLOBINEMIA

The term "methemoglobinemia" or "blue baby syndrome" refers to the disease experienced by an infant who consumes groundwater contaminated with nitrate ions. When an infant consumes formulae made with groundwater contaminated with nitrate ions, the ions are easily converted to nitrite ions in the infant's digestive tract. The nitrite ions that enter the infant's circulatory system bond quickly to the iron in the hemoglobin or red blood cells (Figure 1.4).

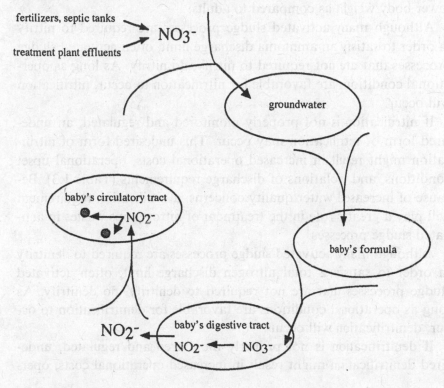

● red blood cell

Figure 1.4 *Methemoglobinemia. If nitrate-contaminated groundwater is used as a potable or drinking water supply, the presence of nitrate ions represents a significant health concern for infants. Nitrate ions may be present in the groundwater due to the overuse of fertilizers, malfunction of septic tanks, or the discharge of high levels of nitrate ions in the effluent of wastewater treatment plants. When an infant consumes nitrate ions from potable water used to prepare baby formula, the nitrate ions are quickly reduced to nitrite ions in the infant's digestive tract. When the nitrite ions enter the infant's circulatory system, they bind quickly and tightly to the iron within the red blood cells or hemoglobin. Once bonded to the red blood cells, oxygen can no longer be transported in adequate quantities throughout the infant's body.*

The presence of nitrite ions on the iron prevents the hemoglobin from obtaining oxygen as it passes through the infant's lungs. The lack of oxygen throughout the infant's body causes the infant's skin to turn blue, thus the term "blue baby syndrome." If insufficient oxygen is present in the infant's brain, paralysis or death may occur.

Methemoglobinemia usually is associated with rural communities where potable water is obtained from groundwater. Methemoglobinemia has no warning sign, and although it can occur in adults, it occurs more rapidly in an infant's due to their lower body pH and lower body weight as compared to adults.

Although many activated sludge processes are required to nitrify in order to satisfy an ammonia discharge limit, often activated sludge processes that are not required to nitrify, do nitrify. As long as operational conditions are favorable for nitrification to occur, nitrification will occur.

If nitrification is not properly monitored and regulated, an undesired form of nitrification may occur. This undesired form of nitrification might result in increased operational costs, operational upset conditions, and violations of discharge requirements (Table 1.3). Because of increased water quality concerns, a nitrification requirement will play a greater role in the treatment of nitrogenous wastes in activated sludge processes.

Although many activated sludge processes are required to denitrify in order to satisfy a total nitrogen discharge limit, often activated sludge processes that are not required to denitrify, do denitrify. As long as operational conditions are favorable for denitrification to occur, denitrification will occur.

If denitrification is not properly monitored and regulated, undesired denitrification might result in increased operational costs, oper-

TABLE 1.3 Operational Problems Associated with an Undesired Form of Nitrification

Operation Problem	Description
Increased operating costs	Increased aeration demand to oxidize NH_4^+ to NO_3^-
	Increased chlorine demand to control filamentous growth
	Increased chlorine demand to control coliform bacteria
Operational upset	Clumping of solids in secondary clarifiers due to denitrification
Permit violation	Interference with effective control of coliform bacteria

TABLE 1.4   Operational Problems Associated with Undesired Denitrification

Operational Problem	Description
Increased operating costs	Increased use of metal salts/polymers to thicken and capture solids in clarifiers
Operational upset	Clumping of solids in secondary clarifiers due to denitrification
Permit violation	Discharge of elevated level of total suspended solids (TSS)

ational upset conditions, and violations of discharge requirements (Table 1.4). Because of increased water quality concerns, a denitrification requirement will play a greater role in the treatment of nitrogenous wastes in activated sludge processes.

Because many activated sludge processes nitrify and denitrify, a review of the microbiology of nitrification and denitrification is desirable for process control, troubleshooting, and cost-effective operation. A review begins with an overview of nitrogen, nitrogenous wastes or compounds, the activated sludge process, and the bacteria involved in the treatment of wastes.

2

The Oxidation States of Nitrogen

The element nitrogen (N) can be found in a large number of organic and inorganic compounds. Organic compounds contain the element carbon (C) and the element (H) (Table 2.1), while inorganic compounds may contain the element carbon or the element hydrogen (Table 2.2). Nitrogen is incorporated into organic compounds and inorganic compounds due to its ability to easily form chemical bonds with other elements such as carbon, hydrogen, and oxygen (O). When elements bond together, compounds are formed.

Organic compounds that contain nitrogen are considered "organic-nitrogen" compounds. An example of an organic-nitrogen compound is urea (NH_2CONH_2). Urea is a major chemical component of urine. Although fresh, domestic wastewater is rich in urea, this compound degrades quickly in the sewer system through bacterial activity. In the sewer system, a large diversity of bacteria adds water to urea. The addition of water to urea "splits" the compounds into ammonium ions and carbon dioxide (CO_2). The "hydrolysis" of urea, namely, the addition of water by bacteria to split a compound into smaller compounds, results in the release of ammonium ions in the sewer system.

Examples of inorganic compounds that contain nitrogen and are of concern to wastewater treatment plant operators include ammonium ions, nitrite ions, and nitrate ions. These ions are the most important nitrogenous compounds that should be monitored for cost-effective operations, permit compliance, and process control for activated sludge processes that are and are not required to nitrify.

TABLE 2.1 Examples of Organic Compounds

Name of the Compound	Chemical Formula of the Compound
Acetic acid	CH_3COOH
Ethyl alcohol	CH_3CH_2OH
Glucose	$C_6H_{12}O_6$
Isopropyl alcohol	$CH_3CHOHCH_3$

All elements are made of atoms that contain a large and positively charged nucleus that is surrounded by small and negatively charged electrons (Figure 2.1). The electrons orbit the nucleus and may be shared between elements. When electrons are shared between elements, chemical bonds and chemical compounds are formed (Figure 2.2). For example, one atom of nitrogen and three atoms of oxygen share electrons. The sharing of electrons forms chemical bonds to produce one molecule of NO_3^-. Although the nitrogen atom and the oxygen atoms shared the electrons, the oxygen atoms pull the electrons more closely to their nuclei. Because the electrons are pulled closely to the oxygen atom, the charge on the nitrogen atom becomes less negative, perhaps positive.

When oxygen is added to nitrogen, the nitrogen atom undergoes oxidation; that is, the nitrogen atom loses electrons. Oxidation of nitrogen results in a decrease in negative charge or an increase in positive charge due to the loss of electrons. Nitrification is the addition of oxygen to nitrogen. Nitrification is the oxidation of nitrogen.

When oxygen is removed from nitrogen, the nitrogen atom undergoes reduction; that is, electrons are returned to the nitrogen atom or the nitrogen atom gains electrons. Reduction of nitrogen results in a decrease in positive charge due to the gain of electrons. Denitrification is the removal of oxygen from nitrogen. Denitrification is the reduction of nitrogen.

TABLE 2.2 Examples of Inorganic Compounds

Name of the Compound	Formula of the Compound
Ammonia	NH_3
Carbon dioxide	CO_2
Copper sulfate	$CuSO_4$
Mercuric chloride	$HgCl_2$

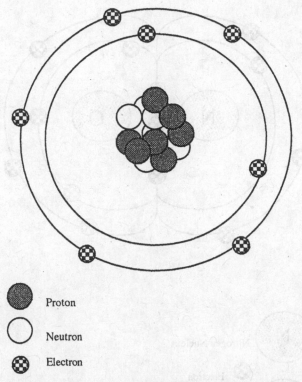

Proton

Neutron

Electron

Figure 2.1 *The atom. The atom contains three basic components, the proton, the neutron, and the electron. The proton and neutron are grouped together at the center or core of the atom and give the atom its weight. The neutron has no charge, while the proton is positively charged. The electron spins around the core of the atom and is negatively charged. If the number of electrons and protons in the atom are equal, the atom has a neutral charge. However, when atoms share electrons, the atoms may become positively or negatively charged depending on the number of shared electrons that spin around the core of each atom.*

Nitrification and denitrification are chemical reactions that occur inside living cells or bacteria. Because these chemical reactions occur in living cells, they are considered "biochemical" reactions.

With all biochemical reactions there are starting compounds or reactants and final compounds or products (Equation 2.1). In some biochemical reactions intermediate compounds may be formed (Equation 2.2). Intermediate compounds usually do not accumulate. However, under appropriate conditions, intermediate compounds may accumulate; for example, nitrite ions are intermediate compounds that may accumulate under appropriate operational conditions. The accumulation of nitrite ions is undesired.

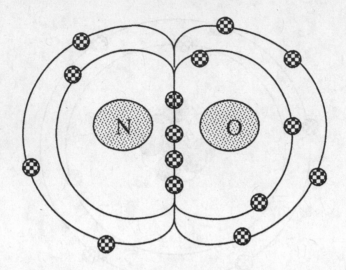

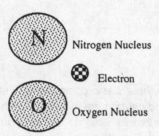

N — Nitrogen Nucleus

— Electron

O — Oxygen Nucleus

Figure 2.2 *Atoms sharing electrons. When a nitrogen atom and an oxygen atom share electrons, the electrons orbit the core of the oxygen atom more than the core of the nitrogen atom. Therefore the nitrogen atom becomes positive in charge (oxidized), while the oxygen atom becomes negative in charge (reduced; that is, its charge is lowered or reduced).*

$$\text{Reactants} \rightarrow \text{Products} \tag{2.1}$$

$$\text{Reactants} \rightarrow \text{Intermediates} \rightarrow \text{Products}$$
$$(NH_4^+ \rightarrow NO_2^- \rightarrow NO_3^-) \tag{2.2}$$

The charge on the nitrogen atom is called its oxidation state or valence, for example, +3 for nitrogen within the ammonium ion. Because changes in the oxidation state of nitrogen can be biologically mediated, that is, caused by biochemical reactions, changes in the oxidation state of nitrogen can occur in the activated sludge process. Therefore nitrogen may undergo nitrification (oxidation) and denitri-

TABLE 2.3   Nitrogenous Compounds Produced during Nitrification and Denitrification

Nitrogenous Compound	Chemical Formula	Oxidation State of N
Nitrate ion	NO_3^-	+5
Nitrite ion	NO_2^-	+3
Nitric oxide	NO	+2
Nitroxyl	NOH	+1
Nitrous oxide	N_2O	+1
Molecular nitrogen	N_2	0
Hydroxylamine	NH_2OH	−1
Ammonia	NH_3	−3
Ammonium ion	NH_4^+	−3

fication (reduction) in an activated sludge process. Numerous oxidation states of nitrogen in the activated sludge process are presented in Table 2.3.

It is the −3 oxidation state of nitrogen in the ammonium ion that is preferred by the bacteria in the activated sludge process as their nitrogen nutrient. In this oxidation state in the ammonium ion, nitrogen is incorporation or assimilated into cellular material ($C_5H_7O_2N$). Nitrate ions and nitrite ions can be used as a nutrient source for nitrogen. However, nitrate ions and nitrite ions are used only after ammonium ions are no longer available and the oxygen on each ion is removed and ammonium ions are produced. The production of ammonium ions inside the bacterial cells ensures the presence of a −3 oxidation state for nitrogen.

3

Nitrogenous Compounds

Because of the many oxidation states of nitrogen, many nitrogenous compounds enter activated sludge processes in domestic wastewater. The diversity of compounds may vary greatly depending on the industrial discharges that contain nitrogenous waste. Examples of nitrogenous compounds that are found in industrial wastewater include analine, chelating agents, corrosion inhibitors, dairy waste, and slaughterhouse waste.

Analine is used in the manufacturing of dyes, photographic chemicals, and drugs. Some chelating agents are organic-nitrogen compounds that are used to hold metals such as copper and iron in solution. Nitrites are used in corrosion inhibitors in industrial process water. Dairy waste contains nitrogen-containing proteins, including casein, and many proteins are present in the meat and blood from slaughterhouse waste.

Domestic wastewater contains organic-nitrogen compounds and ammonium ions. Nitrogen in domestic wastewater originates from protein metabolism in the human body. In fresh domestic wastewater, approximately 60% of the nitrogen is in the organic form, such as proteinaceous wastes, and 40% of the nitrogen is in the inorganic form, such as ammonium ions. Organic compounds such as amino acids, proteins, and urea are the principle organic-nitrogen compounds in domestic wastewater, while ammonium ions are the principle inorganic compound in domestic wastewater.

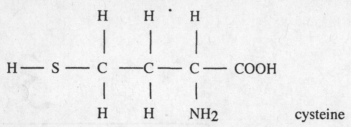

Figure 3.1 *Structure of an amino acid. Regardless of the structure or size of an amino acid, all amino acids contain a carboxyl group (–COOH) and an amino group (–NH₂). In an amino acid such as cyteine, an amino group can be found on the carbon (C) that is bonded to the carboxyl group.*

Unless discharged by specific industries, nitrite ions and nitrate ions are not found in municipal sewer systems. Conditions within the sewer systems are not favorable for the oxidation of ammonium ions or nitrite ions; that is, nitrification does not occur.

AMINO ACIDS

Amino acids are organic-nitrogen compounds that contain the carboxylic acid group (–COOH) and the amino group (–NH₂). The amino group in all amino acids is always bonded to the carbon next to the carboxylic acid group (Figure 3.1). Amino acids are the structural compounds, or building blocks, that form proteins. During bacterial degradation of amino acids, the amino group is released (Figure 3.2).

Deamination is the biochemical reaction responsible for the release of the amino group. Deamination of amino acids can occur in the sewer system and the aeration tank, and deamination can occur in the presence or the absence of dissolved oxygen. Amino acids that are simplistic in structure may be degraded in the sewer system. Amino acids that are complex in structure may be degraded in an aeration tank.

When the amino group is released in wastewater, it is quickly converted to the ammonium ion (Equation 3.1). This conversion occurs in wastewater due to the presence of hydrogen ions (H⁺).

$$NH_2 + 2H^+ \rightarrow NH_4^+ \qquad (3.1)$$

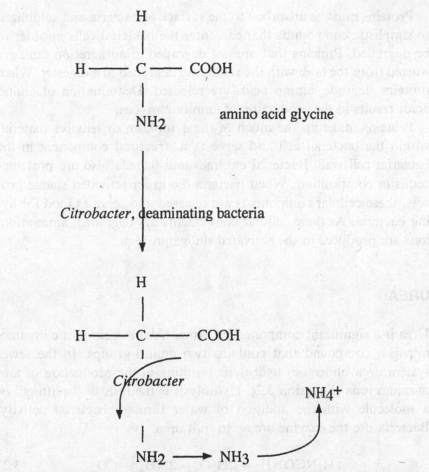

Figure 3.2 *Deamination. When amino acids undergo deamination, the amino groups present on the amino acid are removed by bacterial activity. Here the deaminating bacterium,* Citrobacter, *removes the amino group from the amino acid glycine. Once removed, the amino group is quickly converted to an ammonium ion.*

PROTEINS

Proteins are organic-nitrogen compounds that contain amino acids. Proteins are colloids and are complex in structure. As colloids, they have a large surface area and are suspended in wastewater. Due to their colloidal nature and complex structure, bacterial degradation of proteins is very slow, and deamination usually occurs in an aeration tank containing high concentrations of solids and a very long aeration time.

Proteins must be adsorbed to the surface of bacteria and solublized to simplistic compounds that can enter the bacterial cells in order to be degraded. Proteins that are not degraded in an aeration tank are wasted from the tank with the solids and degraded in a digester. When proteins degrade, amino acids are released. Deamination of amino acids results in the production of ammonium ions.

Proteins make up the much of the cytoplasm or jellylike material within the bacterial cell and serve as a structural component in the bacterial cell wall. Bacterial enzymes and flagella also are proteinaceous in composition. When bacteria die in an activated sludge process, these cellular components are released and serve as food for living bacteria. As these cellular components are degraded, ammonium ions are produced in the activated sludge process.

UREA

Urea is a significant component of urine. Urea is a simplistic organic-nitrogen compound that contains two amino groups. In the sewer system urea undergoes hydrolysis resulting in the production of ammonium ions (Equation 3.2). Hydrolysis is the lysis or "splitting" or a molecule with the addition of water through bacterial activity. Bacteria use the enzyme urease to split urea.

$$H_2NCONH_2 + 2H_2O \rightarrow 2NH_4^+ + CO_2 \qquad (3.2)$$

In the sewer system many organic-nitrogen compounds are rapidly hydrolyzed and deaminated. Due to hydrolysis and deamination in the sewer system, the influent concentration of ammonium ions to activated sludge processes receiving domestic wastewater is usually 15 to 30 mg/l. Although the ammonium ion concentration entering the aeration tanks of these processes is relatively high, additional ammonium ions are released in the aeration tanks as complex amino acids, proteins, and additional nitrogenous wastes are degraded.

4

Bacteria

Bacteria are unicellular organisms that grow as individual cells, pairs, groups of four (tetrads), cubes (sarcinae), irregular clusters or clumps, or chains (Figure 4.1). Individual bacterial cells may be spherical (coccus), rod-shaped (bacillus), or helical (spirillum) (Figure 4.2). Most bacteria are 0.5 to 2.5 microns (μm) in diameter or width and 1 to 20 μm in length. Because of the very small size of bacteria, they can be examined only by using a microscope at high-power magnification. Often, a staining technique, such as Gram staining, is used to examine the bacteria during microscopic work. The Gram staining technique is provided in Appendix I.

All bacteria possess a cell wall, a cell membrane, cytoplasm, mesosomes, ribosomes, and inclusions or storage granules (Figure 4.3). The cell wall surrounds the bacterium and gives the organism its stiffness and shape. The cell wall also provides protection and helps to regulate the movement of compounds in and out of the cell. The cell membrane is a very thin and flexible structure located immediately beneath the cell wall. The cell membrane also helps to regulate the movement of compounds in and out of the cell. Invaginations of the cell membrane are the mesosomes. The invaginations may take the shape of tubules, vesicles, or lamellae. The function of the mesosomes is unknown.

The cytoplasm fills the interior of the cell and is surrounded by the cell membrane. The cytoplasm makes up the bulk content of the cell. It contains a variety of colloids and fluids as well as storage granules

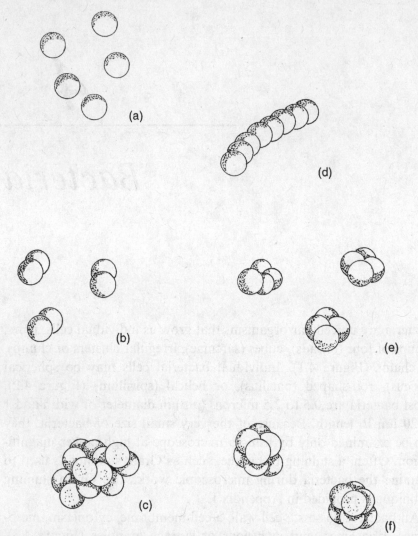

Figure 4.1 *Patterns of bacterial growth. There are several common patterns of growth for bacteria. These patterns include individual (a), pairs (b), irregular clusters (c), chains or filamentous (d), groups of four or tetrads (e), and cubes or sarcinae (f).*

or inclusions. The cytoplasm also contains the mitochondria and ribosomes. The mitochondria are the sites where substrate (food) is degraded, while the ribosomes are the sites where protein synthesis occurs. Inclusions or storage granules consist of food, oils, polyphosphates, and sulfur.

Some bacteria contain a capsule and flagellum (Figure 4.4). The capsule is a gelatinous slime and provides additional protection for

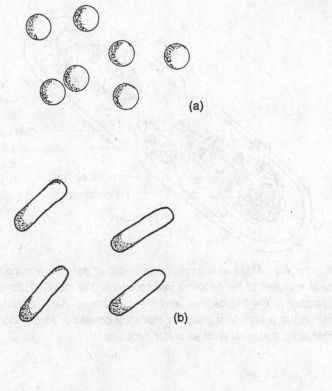

(a)

(b)

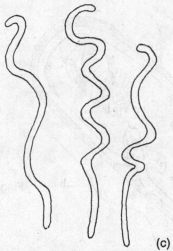

(c)

Figure 4.2 *Shapes of bacterial cells. Most bacterial cells have one of three shapes. These shapes are spherical or coccus (a), rod-shaped or bacillus (b), and helical or spirillum (c). The helical shape is the least rigid of the three bacterial shapes.*

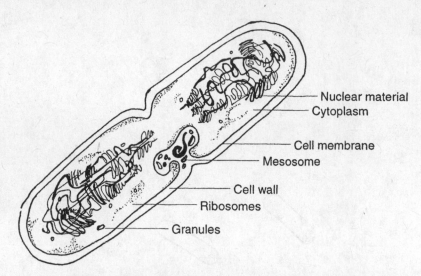

Nuclear material
Cytoplasm

Cell membrane
Mesosome

Cell wall
Ribosomes

Granules

Figure 4.3   *Major structural components of the bacterial cell. Significant, structural features of the bacterial cell consist of the cell wall, the cell membrane, the cytoplasm, the mesosome, and the ribosomes. Also contained within the cytoplasm are a variety of granules that may consist of stored food such as starch or inorganic materials such as sulfur deposits.*

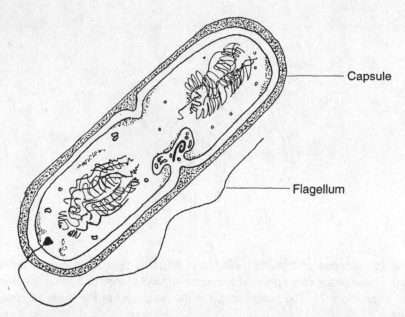

Capsule

Flagellum

Figure 4.4   *Bacterial capsule and flagellum. In addition to the major structural components of the cell, some bacteria may possess a capsule and a flagellum. The capsule provides additional protection for the cell, while the flagellum provides locomotion.*

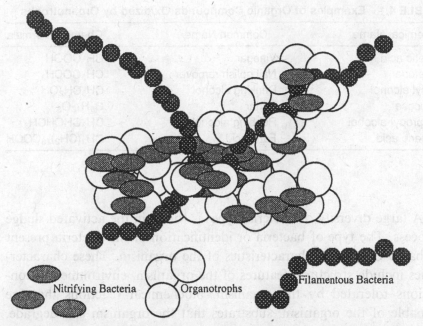

Nitrifying Bacteria   Organotrophs   Filamentous Bacteria

Figure 4.5 *The floc particle. The basis structure of a floc particle consists of numerous bacteria that "stick" together or agglutinate. The bacteria consist of floc-formers such as organotrophs, non-floc formers such as nitrifying bacteria, and filamentous bacteria. The non-floc formers are adsorbed to the floc particle through the coating action from secretions by ciliated protozoa and other higher life forms. Filamentous bacteria grow within the floc particle and extend from the perimeter of the floc particle into the bulk solution. The filamentous bacteria provide the floc particle with strength and permit the floc particle to grow in size.*

the cell. The slime helps to flocculate bacterial cells together, and it also holds particulate and colloidal wastes to the surface of the bacterium.

The flagellum is a proteinaceous, whiplike structure that provides locomotion for the cell. Most bacteria are highly motile when they are young. This locomotive ability is lost when bacteria are incorporated into floc particles (Figure 4.5) or the flagellum is lost through aging.

Bacteria enter the activated sludge process through inflow and infiltration (I/I) as soil and water organisms and as fecal organisms in domestic wastewater. Bacteria in the activated sludge process may be suspended in the water or bulk solution surrounding the floc particles or incorporated into floc particles. Bacteria are present in the activated sludge process at relatively high numbers. They are commonly found in millions per milliliter of bulk solution or billions per gram of floc particles or solids.

TABLE 4.1   Examples of Organic Compounds Oxidized by Organotrophs

Chemical Name	Common Name	Chemical Formula
Acetic acid	Vinegar	CH_3COOH
Acetone	Nail polish remover	CH_3COCH_3
Ethyl alcohol	Drinking alcohol	CH_3CH_2OH
Glucose	Sugar	$C_6H_{12}O_6$
Isopropyl alcohol	Rubbing alcohol	$CH_3CHOHCH_3$
Stearic acid	Fatty acid	$CH_3(CH_2)_{16}COOH$

A large diversity of bacterial types is found in the activated sludge process. The type of bacteria or identification of the bacteria present is based on several characteristics of the organism. These characteristics include structural features of the organism, environmental conditions tolerated by the organism, biochemical reactions that are capable of the organism, substrates that the organism can degrade, and what molecule, such as O_2, is used by the organism to degrade its substrate. The last two characteristics of bacterial identification are of critical importance in a review of nitrification and denitrification and the successful operation of an activated sludge process.

Substrate refers to the food used by bacteria to obtain carbon and energy for cellular activity and cellular growth or reproduction. The degradation of substrate to obtain carbon and energy is cellular respiration.

Bacteria that use organic compounds or carbonaceous wastes to obtain carbon and energy are organotrophs. The term "organo" implies organic, while the term "troph" implies nourishment. Organotrophic bacteria feed upon organic compounds to obtain carbon and energy (Table 4.1). When bacteria use organic compounds to obtain carbon and energy, these compounds or substrates are degraded or oxidized. When these compounds are oxidized in an activated sludge process, the concentration of carbonaceous waste is decreased.

Bacteria that use minerals or inorganic compounds or wastes to obtain energy for cellular activity and cellular growth or reproduction are chemolithotrophs. The term "chemo" refers to chemicals, and the term "litho" refers to minerals. Therefore chemolithotrophs obtain their energy nourishment from chemical minerals. Examples of minerals degraded or oxidized by chemolithotrophs include iron by iron-oxidizing bacteria, sulfur by sulfur-oxidizing bacteria, and nitrogen by nitrifying bacteria (Table 4.2). The bacteria that oxidize inorganic

TABLE 4.2 Examples of Minerals or Inorganic Compounds Oxidized by Chemolithotrophs

Mineral	Compound Containing the Mineral	Chemical Formula of the Compound	Biochemcial (Oxidative) Reaction	Oxidizing Bacteria
Nitrogen	Ammonium ion	NH_4^+	$NH_4^+ + 1.5O_2 \rightarrow$ $NO_2^- + H_2O + 2H^+$	*Nitrosomonas*
Nitrogen	Nitrite ion	NO_2^-	$NO_2^- + 0.5O_2 \rightarrow NO_3^-$	*Nitrobacter*
Sulfur	Sulfite ion	SO_3^{2-}	$SO_3^{2-} + 1.5O_2 + 4H^+$ $\rightarrow SO_4^{2-} + 2H_2O$	*Sulfolobus*

minerals are essential in the recycling of several critical elements between living and nonliving components of the environment.

Some chemolithotrophs, including the bacteria that oxidize nitrogen in the form of ammonium ions and nitrite ions, obtain their carbon from inorganic carbon. The inorganic-carbon source for these organisms is carbon dioxide. When carbon dioxide dissolves in wastewater, carbonic acid (H_2CO_3) is formed (Equation 4.1). In wastewater some of the carbonic acid disassociates and forms the bicarbonate ion (HCO_3^-) and the hydrogen ion (Equation 4.2). It is the bicarbonate ion that is used as the inorganic carbon source.

$$CO_2 + H_2O \rightarrow H_2CO_3 \qquad (4.1)$$

$$H_2CO_3 \leftrightarrow H^+ + HCO_3^- \qquad (4.2)$$

When organisms, like green plants, use inorganic carbon or carbon dioxide, they are referred to as autotrophs. The term "auto" refers to self. Therefore organisms that obtain carbon from carbon dioxide are self-nourishing. Bacteria that obtain carbon from carbon dioxide and energy from the oxidation of chemical minerals are referred to as chemolithoautotrophs.

When bacteria degrade substrate to obtain energy, the chemical bonds in the organic and inorganic compounds are broken and electrons are released (Figure 4.6). The bacteria capture energy from the released electrons and stored it in the form of high-energy, phosphate bonds in the molecule adenosine triphosphate (ATP). After energy has been captured from the released electrons, the electrons must be removed from the cell. A molecule called the final, electron carrier molecule, removes the electrons from the cell (Table 4.3).

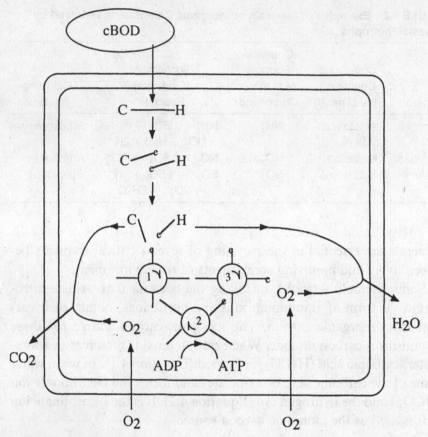

Figure 4.6 *Energy capture by bacterial cell. When the bacterial cell absorbs soluble cBOD, enzymes within the cell split the chemical bonds within the cBOD molecules to release and capture energy. Here the chemical bond between carbon (C) and hydrogen (H) is broken, and two electrons from the bond are released. The released electrons are quickly captured by a series of cellular molecules (1, 2, and 3) that efficiently transport the electrons to an oxygen molecule that carries the electrons out of the cell. As the electrons are passed from one cellular molecule to another, some of the energy from the captured electrons is used to made high-energy phosphates bonds (ADP to ATP). Wastes from the degradation of the cBOD consist of carbon dioxide (CO$_2$) and water (H$_2$O). The oxygen is used to carry waste and electrons from the cell.*

The carrier molecule may be dissolved oxygen, nitrite ions, or nitrate ions. The molecule that is used by the bacterium in the cell in to degrade substrate determines the type of respiration that occurs in the cell. Degradation of substrate that occurs with the use of dissolved oxygen is aerobic respiration. For aerobic respiration to occur, the environment of the bacteria must contain dissolved oxygen; that

TABLE 4.3 Final Electron Carrier Molecules and Types of Respiration

Carrier Molecule	Respiration	Bacterial Environment	Example of Respiration
O_2	Aerobic	Oxic	Nitrification
NO_2^-	Anaerobic	Anoxic	Denitrification
NO_3^-	Anaerobic	Anoxic	Denitrification

is, the environment is "oxic." Degradation of substrate that occurs without the use of free molecular oxygen is anaerobic respiration.

For anaerobic respiration to occur, the environment of the bacteria must contain no dissolved oxygen; that is, the environment is anaerobic. Denitrification is one form of anaerobic respiration. During denitrification nitrite ions or nitrate ions are used to degrade substrate. Therefore the environment of the bacteria must contain nitrite ions or nitrate ions. This environment is referred to as an "anoxic" environment.

5

The Activated Sludge Process

The activated sludge process is the most commonly used system for the treatment of municipal wastewater, and it is probably the most versatile and effective of all wastewater treatment processes. The treatment of wastes is biological in that it uses microscopic organisms to degrade or remove wastes. The process consists of at least one aeration tank and one clarifier (Figure 5.1).

Often a sedimentation tank or clarifier is placed upstream of the activated sludge process. The purpose of this first or primary clarifier is to remove floating materials such as oils and greases and heavy solids that settle to the bottom of the clarifier. If a primary clarifier is placed upstream of the activated sludge process, the clarifier following the aeration tank is secondary clarifier.

The aeration tank is a biological reactor or amplifier where relatively large numbers of bacteria are provided with dissolved oxygen and carbonaceous and nitrogenous wastes. In the presence of dissolved oxygen, the bacteria degrade the carbonaceous and nitrogenous wastes. The degradation of the wastes by the bacteria (biological reactor) results in the growth of the bacterial population (biological amplifier).

The wastes that are degraded by the bacteria are the substrates used to obtain carbon and energy. The term given for the substrates is biochemical oxygen demand (BOD). The BOD is the amount of dissolved oxygen measured in milligrams per liter (mg/l) required by the organisms, primarily bacteria, to oxidize (degrade) the wastes to simple inorganic compounds and more bacterial cells.

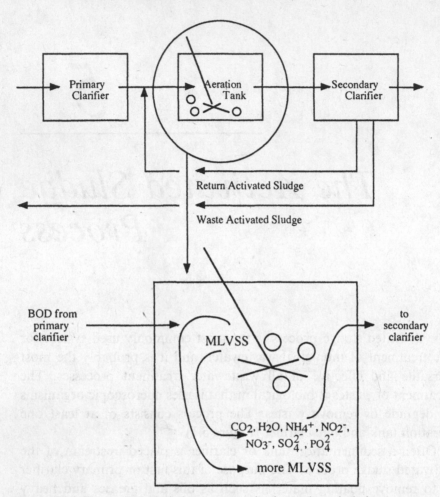

Figure 5.1 The activated sludge process. In the activated sludge process, a tank (aeration tank) or series of tanks is used to degrade or oxidize wastes. The oxidation of the wastes is achieved by mixing the wastes, with air and a high concentration of bacteria, for sufficient time. When the wastes are oxidized more bacteria are produced. The bacteria in the form of floc particles leave the aeration tank and enter a clarifier. Here the floc particles settle to the bottom of the clarifier and are returned to the aeration tank to degrade more incoming wastes or are removed from the process. If a clarifier is placed upstream of the aeration tank, the clarifier is called a "primary clarifier." The purpose of this clarifier is to remove heavy solids and fats, oils, and grease. The clarifier following the aeration tank usually is called a "secondary clarifier." Wastes oxidized in the aeration tank are converted to carbon dioxide (CO_2), water (H_2O), ammonium ions (NH_4^+), nitrite ions (NO_2^-), nitrate ions (NO_3^-), sulfate ions (SO_4^{2-}), phosphate ions (PO_4^{2-}), and more bacterial cells (MLVSS).

TABLE 5.1 Inorganic Products Released in the Aeration Tank from the Oxidation of Proteins

Element Contained in Proteins	Inorganic Product Released
Carbon	Carbon dioxide (CO_2)
Hydrogen	Water (H_2O)
Nitrogen	Ammonium ion (NH_4^+)
Oxygen	Water (H_2O)
Phosphorus	Phosphate ion (PO_4^{2-})
Sulfur	Sulfate ion (SO_4^{2-})

Ultimately, under appropriate operational conditions and adequate aeration time, the bacteria convert the substrate to simplistic products through biochemical reactions. Some organic-nitrogen compounds such as proteins contain carbon, hydrogen, nitrogen, oxygen, phosphorus, and sulfur. The sulfur is found in thiol groups (–SH) bonded to the proteins. When proteins are degraded in the aeration tank by bacteria, the bacteria obtain carbon for growth, energy for cellular activity, and release inorganic products in the aeration tank (Table 5.1).

Ammonium ions that are produced in the sewer system and the aeration tank through hydrolysis and deamination are the substrate for the bacteria that oxidize nitrogen in the form of ammonium ions. The oxidation of the ammonium ions by bacteria is nitrification. When ammonium ions are oxidized, the bacteria obtain energy and release nitrite ions in the aeration tank.

The nitrite ions that are produced in the aeration tank are the substrate for the bacteria that oxidize nitrogen in the form of nitrite ions. The oxidation of nitrite ions by bacteria is nitrification. When nitrite ions are oxidized, the bacteria obtain energy and release nitrate ions in the aeration tank.

When bacterial cells oxidize substrate in the aeration tank, reproduction occurs or an increase in the bacterial population results. Bacteria represent a portion of the solids in the aeration tank. Therefore, as the bacterial population increases through reproduction, the solids inventory in the aeration tank also increases.

Solids in the aeration tank are referred to as sludge. Because the sludge is aerated, and the bacteria become very active during aeration, the term "activated sludge" is used to describe the process where bacterial solids are active in the purification of the wastes within the aeration tank.

As the bacteria in the aeration tank age, many bacteria stick together to form floc particles or large solids. These particles contain a large number and diversity of bacteria that degrade the wastes in the aeration tank. As the solids flow into the secondary clarifier, the solids settle to the floor of the secondary clarifier and a clear supernatant develops above the settled solids. After additional treatment, the supernatant is discharged to the receiving water.

The settled solids may be returned to the aeration tank or may be removed from the activated sludge process. The removal of solids from the activated sludge system is referred to as wasting. Solids are "wasted" to another treatment unit for additional treatment and disposal.

There are several operational factors that can be used to monitor and regulate the activated sludge process. These factors include F/M and MCRT (Appendix II). These factors are critical for monitoring and regulating nitrification and denitrification in the activated sludge process.

F/M is the food-to-microorganism ratio. This factor measures the quantity of BOD or food (the "F" in F/M) available per day per quantity of bacteria or microorganisms (the "M" in F/M). The F/M increases when additional food enters the activated sludge process or more bacteria are wasted from an activated sludge process. The F/M decreases when less food enters an activated sludge process or fewer bacteria are wasted from an activated sludge process.

The MCRT is the mean cell residence time as measured in days. The MCRT is the average time the solids or bacteria are retained in an activated sludge process. The higher the MCRT is the older the bacteria are.

The MCRT is increased in an activated sludge process by decreasing the quantity of solids wasted. The MCRT is decreased in an activated sludge process by increasing the quantity of solids wasted.

Part II

Nitrification

6

Introduction to Nitrification

Biological nitrification is the conversion or oxidation of ammonium ions to nitrite ions and then to nitrate ions. During the oxidation of ammonium ions and nitrite ions, oxygen is added to the ions by a unique group of organisms, the nitrifying bacteria (Figure 6.1). Nitrification occurs in nature and in activated sludge processes. Nitrification in soil is especially important in nature, because nitrogen in absorbed by plants as a nutrient in the form of nitrate ions. Nitrification in water is of concern in wastewater treatment, because nitrification may be required for regulatory purposes or may contribute to operational problems.

Although ammonium ions and ammonia are reduced forms of nitrogen, that is, are not bonded to oxygen, it is the ammonium ion, not ammonia, that is oxidized during nitrification. The quantities of ammonium ions and ammonia in an aeration tank are dependent on the pH and temperature of the activated sludge (Figure 6.2). In the temperature range of 10° to 20°C and pH range of 7 to 8.5, which are typical of most activated sludge processes, about 95% of the reduced form of nitrogen is present as ammonium ions.

The oxidation of ammonium ions and nitrite ions is achieved through the addition of dissolved oxygen within bacterial cells. Because nitrification or the biochemical reactions of oxygen addition occur inside biological cells, nitrification occurs through biochemical reactions.

The ammonium ions are produced in the wastewater from the

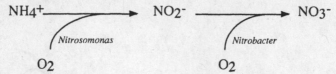

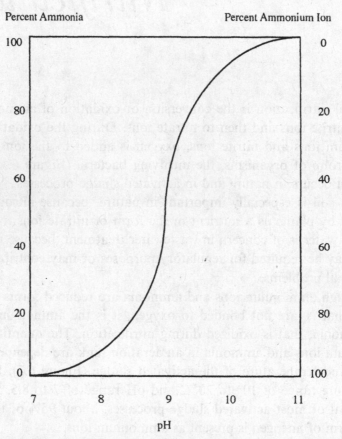

Figure 6.1 *Biological nitrification. Biological nitrification in the activated sludge process consists of the removal of oxygen from an aeration tank and its addition to ammonium ions or nitrite ions. Oxygen is added to ammonium ions by the nitrifying bacterium* Nitrosomonas, *while oxygen is added to nitrite ions by the nitrifying bacterium* Nitrobacter.

Figure 6.2 *pH and the conversion of ammonia and ammonium ions. The relative quantities of ammonia and ammonium ions in wastewater or the activated sludge process are determined by the pH of the wastewater or aeration tank. As the pH of the wastewater decreases, ammonium ions are favored. As the pH of the wastewater increases, ammonia is favored. At a pH value of 9.4 or higher, ammonia is strongly favored.*

**TABLE 6.1 Organisms in the Aeration Tank That
Are Capable of Nitrification**

Organism	Genus
Actinomycetes	*Myocbacterim*
	Nocardia
	Streptomyces
Algae	*Chlorella*
Bacteria	*Arthrobacter*
	Bacillus
	Nitrobacter
	Nitrosomonas
	Proteus
	Pseudomonas
	Vibrio
Fungi	*Aspergillus*
Protozoa	*Epistylis*
	Vorticella

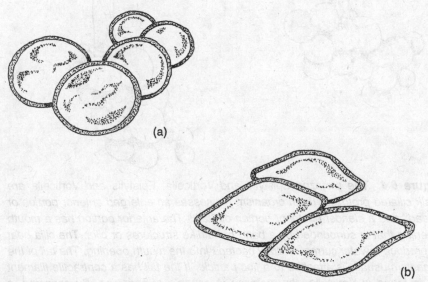

Figure 6.3 *The nitrifying bacteria* Nitrosomonas *and* Nitrobacter. *The nitrifying bacteria of importance in the activated sludge process consist of the coccus-shaped* Nitrosomonas *(a) and the bacillus-shaped* Nitrobacter *(b).* Nitrosomonas *is 0.5 to 1.5 μm in size or diameter, while* Nitrobacter *is approximately 0.5 × 1.0 μm in size.*

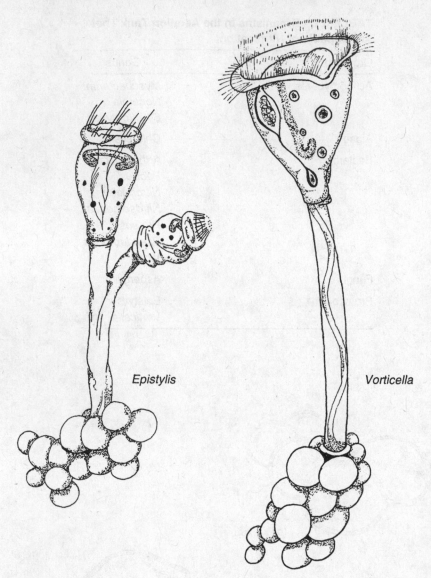

Epistylis

Vorticella

Figure 6.4   *The protozoa* Epistylis *and* Vorticella. Epistylis *and* Vorticella *are stalk ciliated protozoa. Each organism possesses an enlarged anterior portion or "head" and a slender posterior portion or "tail." The anterior portion has a mouth opening that is surrounded by a band of hairlike structures or cilia. The cilia beat to produce a water current to draw bacteria into the mouth opening. The tail of the organism usually is attached to a floc particle. If the tail has a contractile filament as is found in* Vorticella, *the protozoa is capable of springing. By springing, a stronger water current is produced for drawing bacteria into the mouth opening.* Epistylis *is colonial, while* Vorticella *is solitary (single). Species of* Epistylis *vary in size from 50 to 250 μm, while species of* Vorticella *vary in size from 30 to 150 μm.*

hydrolysis of urea and degradation of organic-nitrogen compounds. Hydrolysis and degradation of organic-nitrogen compounds results in the release of amino groups and the production of ammonium ions.

Although there are many organisms that are capable of oxidizing ammonium ions and nitrite ions (Table 6.1), the principle organisms responsible for most, if not all, nitrification in the activated sludge process are two genera of nitrifying bacteria, *Nitrosomonas* and *Nitrobacter* (Figure 6.3). These genera of bacteria possess special enzymes and cellular structures that permit them to achieve significant nitrification.

The genera of nitrifying bacteria that oxidize ammonium ions to nitrite ions are prefixed Nitroso- (such as *Nitrosomonas*), and the genera of nitrifying bacteria that oxidize nitrite ions to nitrate ions are prefixed Nitro (such as *Nitrobacter*). Nitrification by organisms other than nitrifying bacteria occurs at relatively low rates, and it is not associated with cellular growth or reproduction.

The rate of nitrification achieved by nitrifying bacteria is often 1000 to 10,000 times greater than the rate of nitrification achieved by other organisms. Besides the nitrifying bacteria, there are two protozoa that are present in relatively large numbers during rapid nitrification. These protozoa are *Epistylis* and *Vorticella* (Figure 6.4). However, it is not known if these protozoa are capable of a rapid rate of nitrification, or if they simple grow in large numbers under operational conditions that are optimal for nitrification to occur through nitrifying bacteria.

Although activated sludge processes are used for nitrification, these processes are not ideal for nitrification. Due to the large population size and rapid growth of organotrophs in the aeration tank as compared to the small population size and slow growth of nitrifying bacteria, the population size of nitrifying bacteria is gradually diluted, making it difficult to achieve and maintain desired nitrification. Approximately 90% to 97% of the bacteria in the activated sludge process consists of organotrophs, while the remaining 3% to 10% of the bacteria are nitrifiers.

7

Nitrifying Bacteria

Nitrifying bacteria live in large variety of habitats including freshwater, potable water, wastewater, marine water, brackish water, and soil. Nitrifying bacteria are known by many names that are derived from the carbon and energy substrates (Table 7.1).

Although some genera of nitrifying bacteria are capable of using some organic compounds to obtain carbon, the principal genera of nitrifying bacteria in the activated sludge process, *Nitrosomonas* and *Nitrobacter*, use carbon dioxide or inorganic carbon as their carbon source for the synthesis of cellular material. For each molecule of carbon dioxide assimilated into cellular material by nitrifying bacteria, approximately 30 molecules of ammonium ions or 100 molecules of nitrite ions must be oxidized.

Due to the relatively large quantity of ammonium ions and nitrite ions needed to assimilate carbon dioxide, nitrifying bacteria have a very low reproductive rate. Even under the best conditions the reproductive rate of nitrifying bacteria is minimal. In the activated sludge process, nitrifying bacteria are able to increase in number only if their reproductive rate is greater than their removal rate through sludge wasting and discharge in the final effluent. Therefore a high MCRT is required to increase the number of nitrifying bacteria in the activated sludge process.

For the growth of one pound of dry cells, *Nitrosomonas* must oxidize 30 pounds of ammonium ions, while *Nitrobacter* must oxidize 100 pounds of nitrite ions. In contrast, for the growth of one pound

TABLE 7.1   Common Names of Nitrifying Bacteria

Name	Derivation
Ammonia-oxidizing bacteria	Oxidize ammonium ions
Ammonia-removing bacteria	Reduce the concentration of ammonium ions
Autotrophs	Obtain carbon from CO_2
Chemolithoautotrophs	Obtain carbon from CO_2 and energy from chemical minerals
nBOD-oxidizing bacteria	Oxidize ammonium ions Oxidize nitrite ions
nBOD-removing bacteria	Reduce the concentration of ammonium ions Reduce the concentration of nitrite ions
Nitrifiers	Oxidize ammonium ions Oxidize nitrite ions
Nitrifying bacteria	Oxidize ammonium ions Oxidize nitrite ions
Nitrite-oxidizing bacteria	Oxidize nitrite ions
Nitrite-removing bacteria	Reduce the concentration of nitrite ions
NOD-oxidizing bacteria	Oxidize ammonium ions Oxidize nitrite ions
NOD-removing bacteria	Reduce the concentration of ammonium ions Reduce the concentration of nitrite ions

of dry cells, the organotrophic bacterium *Escherichia coli* must oxidize only two pounds of glucose.

Nitrifying bacteria obtain their energy by oxidizing inorganic substrates, namely ammonium ions and nitrite ions (Table 7.2). Nitrite ions that are the product of the oxidation of ammonium ions by *Nitrosomonas* serve as the substrate for *Nitrobacter* (Figure 7.1). Unless nitrite ions are discharged to the activated sludge process, nitrite ions must be produced in the aeration tank in order for *Nitrobacter* to have an energy substrate.

TABLE 7.2   Oxidation Reactions of Nitrifying Bacteria

Oxidation Reaction	Genus Responsible for the Oxidation Reaction
$NH_4^+ + 1.5O_2 \rightarrow NO_2^- + H_2O + 2H^+$	*Nitrosomonas*
$NO_2^- + 0.5O_2 \rightarrow NO_3^-$	*Nitrobacter*

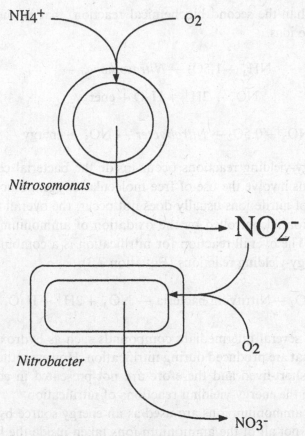

Figure 7.1 *Significant waste product of* Nitrosomonas. *Nitrite ions are the significant waste product of* Nitrosomonas. *These ions are important for two reasons. First, unless nitrite ions are discharged to the activated sludge process by an industry,* Nitrobacter *must wait for* Nitrosomonas *to oxidized ammonium ions and then release nitrite ions to the bulk solution in order to have an energy substrate. Second, unless* Nitrobacter *oxidizes nitrite ions, these ions would accumulate in the activated sludge process and would interfere with the ability of chlorine to destroy coliform organisms and filamentous organisms.*

Nitrifying bacteria belong in the family Nitrobacteracae. With some exceptions, bacteria in this family obtain carbon by assimilating carbon dioxide to a 5-carbon sugar, ribulose diphosphate, to produce a 6-carbon sugar, glucose. The assimilation of carbon dioxide results in the production of cellular material. When carbon dioxide is assimilated, the carbon dioxide is reduced by the addition of hydrogen.

There are two energy-yielding, biochemical reactions that occur during nitrification (Equations 7.1 and 7.2). More energy is derived from the first biochemical reaction, that is, the oxidation of ammonium ions, than the second biochemical reaction, that is, the oxidation of nitrite ions.

$$NH_4^+ + 1.5O_2 - Nitrosomonas \rightarrow$$
$$NO_2^- + 2H^+ + H_2O + energy \qquad (7.1)$$

$$NO_2^- + 0.5O_2 - Nitrobacter \rightarrow NO_3^- + energy \qquad (7.2)$$

The energy-yielding reactions occur inside the bacterial cells, and both reactions involve the use of free molecular oxygen. Since an accumulation of nitrite ions usually does not occur, the overall reaction for nitrification is controlled by the oxidation of ammonium ions to nitrite ions. The overall reaction for nitrification is a combination of the two energy-yielding reactions (Equation 7.3).

$$NH_4^+ + 2O_2 - Nitrifying\ bacteria \rightarrow NO_3^- + 2H^+ + H_2O \qquad (7.3)$$

There are several intermediate compounds such as hydroxylamine (NH_2OH) that are produced during nitrification. However, these compounds are short-lived and therefore are not presented in equations that describe the energy-yielding reactions of nitrification.

Although ammonium ions are used as an energy source by nitrifying bacteria, not all of the ammonium ions taken inside the bacterial cells are nitrified. Some of the ammonium ions are used as a nutrient source for nitrogen and are assimilated into new cellular material ($C_5H_7O_2N$). The growth of new cells in the activated sludge process is referred to as an increase in the mixed liquor volatile suspended solids (MLVSS) (Equation 7.4).

$$4CO_2 + HCO_3^- + NH_4^+ + 4H_2O \rightarrow C_5H_7O_2N + 5O_2 + 3H_2O \qquad (7.4)$$

Carbon dioxide serves as the carbon source for the synthesis of cellular material and is made available to nitrifying bacteria as bicarbonate alkalinity. This alkalinity is produced when carbon dioxide dissolves in wastewater.

There are several genera of nitrifying bacteria. The genera can be grouped according to those that oxidize ammonium ions and those that oxidize nitrite ions (Table 7.3).

Nitrifying bacteria are not pathogenic (disease-causing) and are not common or indigenous to the intestinal tract of humans. Therefore nitrifying bacteria do not enter sewer systems and activated sludge

TABLE 7.3 Genera of Nitrifying Bacteria

Energy Substrate	Oxidized Product	Genera of Nitrifying Bacteria
NH_4^+	NO_2^-	Nitrosococcus
		Nitrosocystis
		Nitrosolobus
		Nitrosomonas
		Nitrosospira
NO_2^-	NO_3^-	Nitrobacter
		Nitrococcus
		Nitrospira

processes in large numbers through domestic wastewater. Nitrifying bacteria are indigenous to soil and water. Therefore large numbers of nitrifying bacteria enter sewer systems and activated sludge processes through I/I.

The nitrifying bacteria *Nitrosomonas* and *Nitrobacter* are largely, if not entirely, responsible for nitrification in soil. Because nitrifying bacteria are destroyed by ultraviolet light, they are not found in large numbers on the surface of the soil. However, they are found in large numbers immediately beneath the surface of the soil where ultraviolet light cannot penetrate.

Activated sludge processes are seeded with *Nitrosomonas* and *Nitrobacter* through I/I, and these genera of nitrifying bacteria also are largely, if not entirely, responsible for nitrification in the activated sludge process. The principle species of nitrifying bacteria responsible for the oxidation of ammonium ions and nitrite ions are *Nitrosomonas europeae* and *Nitrobacter agilis*, respectively. Other genera of nitrifying bacteria are of secondary importance in the activated sludge process.

Nitrosomonas and *Nitrobacter* are Gram-negative bacteria and are strict aerobes that require free molecular oxygen or dissolved oxygen in order to oxidize substrate. Gram-negative bacteria stain red when dried smears of the bacteria are exposed to a series of dyes. Oxygen is used to carry and remove electrons from the bacterial cell as they are released during the oxidation of ammonium ions and nitrite ions.

Although nitrifying bacteria can grow and reproduce in the presence of most organic compounds, some simplistic forms of organic compounds can inhibit their activity, that is, inhibit nitrification. These inhibitory compounds include alcohols and acids. Some organic com-

TABLE 7.4 Basic Physiological and Structural Features of *Nitrosomonas* and *Nitrobacter*

Feature	*Nitrosomonas*	*Nitrobacter*
Carbon source	Inorganic (CO_2)	Inorganic (CO_2)
Cell shape	Coccus (spherical)	Bacillus (rod-shaped)
Cell size, μm	0.5 to 1.5	0.5×1.0
Habitat	Soil and water	Soil and water
Motility	Yes	No
Oxygen requirement	Strict aerobe	Strict aerobe
pH growth range	5.8 to 8.5	6.5 to 8.5
Reproduction mode	Binary fission	Budding
Generation time	8 to 36 hours	12 to 60 hours
Temperature growth range	5° to 30 °C	5° to 40 °C
Sludge yield	0.04 to 0.13 pound of cells per # of NH_4^+ oxidized	0.02 to 0.07 pound of cells per # of NO_2^- oxidized
Cytomembranes	Present	Present

pounds that have amino groups, such as methylamine (CH_2NH_2), also inhibit the activity of nitrifying bacteria.

With some exceptions, nitrifying bacteria are obligate (strict) autotrophs. Because they are obligate autotrophs, some simplistic forms of organic compounds that remain in the aeration tank inhibit nitrifying bacteria, that is, inhibit nitrification. Therefore a large and diverse population of organotrophs must be present in the aeration tank in order to oxidize these simplistic forms of organic compounds.

Basic physiological and structural characteristics of *Nitrosomonas* and *Nitrobacter* are provided in Table 7.4. Important structural features listed in Table 7.4 are the cytomembrane.

The cytomembranes of nitrifying bacteria intrude from the cell membrane into the peripheral or inner region of the cell (Figure 7.2). The cytomembranes are the active site for the oxidation of ammonium ions and nitrite ions. Here the oxidation of the ions occurs at relatively high rates.

In an activated sludge process the oxidation of ammonium ions and nitrite ions consists of their adsorption to the surface of the cytomembranes and their oxidation by enzymes on the surface of the cytomembranes. After these ions have been oxidized, their respective waste or product (nitrite ions and nitrate ions) are released to the bulk solution (Figure 7.3). As the ions are oxidized, energy is obtained and stored in the form of high-energy, phosphate bonds.

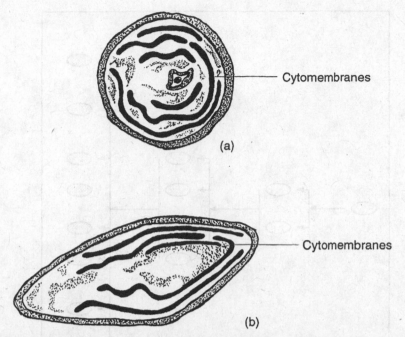

Figure 7.2 *Cytomembranes of nitrifying bacteria. The cytomembranes of the nitrifying bacteria* Nitrosomonas *(a) and* Nitrobacter *(b) are the active sites for the nitrification of ammonium ions and nitrite ions. The cytomembranes consist of extensions or an accordionlike folding of the cell membrane away from the cell wall and toward the center of the cytoplasm.*

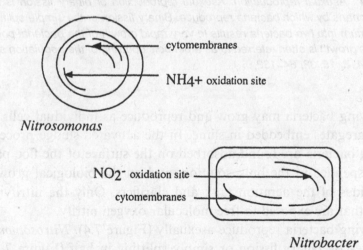

Figure 7.3 *Oxidation of ammonium ions and nitrite ions on the cytomembranes. It is on the cytomembranes of* Nitrosomonas *and* Nitrobacter *where ammonium ions and nitrite ions, respectively, come in contact with the enzymes that add oxygen to each ion. Ammonium ions removed from the bulk solution are oxidized on the cytomembranes of* Nitrosomonas, *while nitrite ions removed from the bulk solution are oxidized on the cytomembranes of* Nitrobacter.

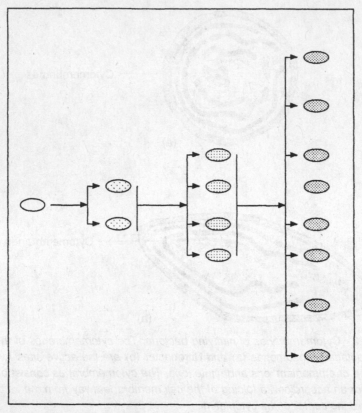

Figure 7.4 *Asexual reproduction. Asexual reproduction or binary fission is the principle means by which bacteria reproduce. Binary fission or the simple splitting of a bacterium into two bacteria results in very rapid growth of the bacterial population. This growth is often referred to as a periodic doubling of the population size (i.e., 1, 2, 4, 8, 16, 32, 64, 132,...).*

Nitrifying bacteria may grow and reproduce as individual cells or small aggregates embedded in slime. In the activated sludge process, nitrifying bacteria are found adsorbed on the surface of the floc particles, suspended in the bulk solution, and on the biological growth on the sides of the aeration tank and clarifiers. Only the nitrifying bacteria that are exposed to free molecular oxygen nitrify.

Nitrifying bacteria reproduce asexually (Figure 7.4). *Nitrosomonas* reproduces by binary fission or simply splitting in half (Figure 7.5), while *Nitrobacter* reproduces by budding (Figure 7.6).

Although nitrifying bacteria are poor floc-forming bacteria, they are incorporated in the floc particles through two means. First, nitrifying bacteria that possess a compatible charge are quickly removed

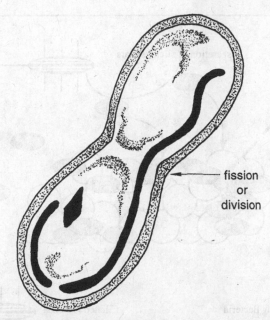

Figure 7.5 *Binary fission in* Nitrosomonas. *When* Nitrosomonas *reproduces, the cell simply divides into two cells. This division occurs as the cell is "pinched" in at the center into two new cells.*

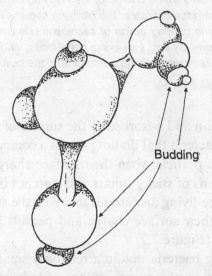

Figure 7.6 *Budding in* Nitrobacter. *Like yeast cells,* Nitrobacter *reproduces by budding. When budding occurs, new cellular material from the original or parent cells "pushes" free to form a new or daughter cell.*

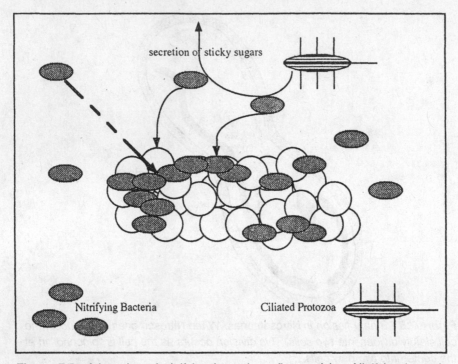

secretion of sticky sugars

Nitrifying Bacteria

Ciliated Protozoa

Figure 7.7 *Adsorption of nitrifying bacteria to floc particles. Nitrifying bacteria are poor floc-forming bacteria and are adsorbed to the surface of floc particles through two mechanisms. First, if the surface charge of the nitrifying bacteria is compatible for adsorption, the nitrifying bacteria are added directly to the growing floc particle. Second, if the surface charge of the nitrifying bacteria is not compatible for adsorption, the surface charge of the nitrifying bacteria is made compatible for adsorption through the coating action of secretions released by ciliated protozoa. The secretions from rotifers, free-living nematodes, and other multicellular organisms also help to remove nitrifying bacteria from the bulk solution to the surface of floc particles.*

from the bulk solution and adsorbed to the surface of the floc particle. Second, nitrifying bacteria that do not possess a compatible charge are adsorbed to the floc particle when their surface charge is made compatible. The secretions of sticky sugars and starches by ciliated protozoa, rotifers, and free-living nematodes that coat the surface of nitrifying bacteria alters their surface charge and permits their adsorption to the floc particles (Figure 7.7).

Because nitrifying bacteria obtain a relatively small amount of energy from the oxidation of ammonium ions and nitrite ions, reproduction or generation time is slow, and a small population of organisms or MLVSS is produced per pound of ammonium ions and nitrite ions oxidized. The population size of nitrifying bacteria within the activated sludge is very small in comparison to the population size of organotrophs.

The difference in population size between nitrifying bacteria and organotrophs is due to two reasons. First, in most municipal and industrial activated sludge processes, the concentration of carbonaceous wastes greatly exceeds the concentration of nitrogenous wastes. Therefore more substrate is available to grow more organotrophs. Second, organotrophs obtain more energy for reproduction than do nitrifying bacteria when they oxidize their respective substrates. Therefore organotrophs can reproduce more quickly than nitrifying bacteria can reproduce.

Compared to organotrophs the generation time of nitrifying bacteria is much longer. The generation time of most organotrophs within the activated sludge process is 15 to 30 minutes. Under favorable conditions the generation time of nitrifying bacteria within the activated sludge process is 48 to 72 hours.

Within the nitrifying bacterial population there also is a difference in population sizes between *Nitrosomonas* and *Nitrobacter*. The population size of *Nitrosomonas* is larger than *Nitrobacter*. Because *Nitrosomonas* obtains more energy from the oxidation of ammonium ions than *Nitrobacter* obtains from the oxidation of nitrite ions, *Nitrosomonas* has a shorter generation time and is able to increase quickly in numbers as compared to *Nitrobacter* (Table 7.5). A larger population size of *Nitrosomonas* than *Nitrobacter* in the activated sludge process provides for more ammonium ion oxidizing ability than nitrite ion oxidizing ability.

The difference in generation times between *Nitrosomonas* and *Nitrobacter* directly affects nitrification. The difference is in part responsible for the buildup of nitrite ions during unfavorable operational conditions including cold temperature, hydraulic washout, low dissolved oxygen level, start-up, toxicity, and slug discharge of soluble cBOD.

In the aeration tank of municipal and industrial activated sludge processes a relatively high MCRT (> 8 days) is required in order to provide an opportunity for the slow growing and poorly flocculating population of nitrifying bacteria to increase in size. Even with a relatively high MCRT, the population size of the nitrifying bacteria is normally less than 10% of the total bacterial population in the aeration tank.

Although the ultimate population size of nitrifying bacteria is dependent on the amount of substrate (ammonium ions and nitrite ions)

TABLE 7.5 Increase in Population Sizes of *Nitrosomonas* and *Nitrobacter* over 72 Hours

Hours	Nitrosomonas		Nitrobacter	
	Divisions	Bacteria	Divisions	Bacteria
0	0	1	0	1
8	1	2	0	1
16	2	4	1	2
24	3	8	2	4
32	4	16	2	4
40	5	32	3	8
48	6	64	4	16
56	7	128	4	16
64	8	256	5	32
72	9	512	6	64

Note: The table assumes that the *Nitrosomonas* population and the *Nitrobacter* population begin with one bacterium each, no bacterium dies, and *Nitrosomonas* and *Nitrobacter* reproduce at optimum generation times in the laboratory, 8 hours for *Nitrosomonas* and 12 hours for *Nitrobacter*. After 72 hours (three days) the population size of *Nitrosomonas* is eight times larger than the population size of *Nitrobacter*. The difference in population sizes over time and the difference in energy obtained from the oxidation of substrates permits ammonium ions to be more easily oxidized than nitrite ions in the activated sludge process.

available, the growth and reproduction of the population is strongly influenced by several operational factors including dissolved oxygen, alkalinity and pH, temperature, inhibition, toxicity, and mode of operation.

For example, the growth rate of nitrifying bacteria is directly influenced by temperature. With increasing temperature, the growth of the nitrifying bacteria accelerates, and nitrification is achieved with little difficulty. With decreasing temperature, the growth of the nitrifying bacteria slows, and nitrification is achieved with much difficulty. Therefore during cold temperatures a buildup of nitrite ions may occur as *Nitrobacter* oxidizes nitrite ions more slowly than *Nitrosomonas* oxidizes ammonium ions.

The presence of nitrifying bacteria in an activated sludge process can be identified by the production of nitrite ions or nitrate ions across the aeration tank. Nitrifying bacteria can be cultivated on selective agar. However, due to the slow generation time of nitrifying bacteria and their poor isolation and colony development on agar media, it is often difficult to identify nitrifying bacteria on selective agar.

8

Organotrophs

In the activated sludge process, numerous organisms, particularly bacteria, work together to oxidize and remove wastes. The successful operation of the activated sludge process involves the management of abundant and active populations of organotrophs and nitrifying bacteria. Proper interactions between organotrophs and nitrifying bacteria are required to remove carbonaceous BOD (cBOD) and nitrogenous BOD (nBOD).

Organotrophs remove soluble cBOD, particulate BOD (pBOD), and colloidal BOD (coBOD). Particulate BOD and colloidal BOD are forms of cBOD. The organotrophs are known by several names (Table 8.1). The names also are derived from their carbon and energy substrates.

Organotrophs often are referred to as heterotrophs because they obtain the carbon they need for cellular growth from organic wastes, and not from carbon dioxide. Because they derive energy from the oxidation of organic wastes, they are called chemotrophs. Therefore organotrophs are chemoorganotrophs; namely they derive carbon and energy from the oxidation of organic compounds. A comparison of carbon and energy substrates of organotrophs and nitrifying bacteria is provided in Table 8.2.

By using organic wastes as a substrate for carbon and energy, organotrophs reduce the quantity of cBOD in the activated sludge process. By using carbon dioxide as a carbon substrate and ammonium ions and nitrite ions as energy substrates, nitrifying bacteria decrease

TABLE 8.1 Common Names of Organotrophs

Name	Derivation
cBOD-oxidizing bacteria	Oxidize organic wastes
cBOD-removing bacteria	Reduce the concentration of organic wastes
Chemoorganotrophs	Obtain carbon and energy from organic wastes
Heterotrophs	Do not obtain carbon from carbon dioxide
Organotrophs	Obtain carbon and energy from organic wastes

alkalinity and pH in the aeration tank and reduce the quantity of nBOD in the activated sludge process.

In order for nBOD to be reduced, cBOD first must be reduced to a relatively low concentration (< 40 mg/l) to ensure that adequate degradation of those soluble and simplistic forms of cBOD that inhibit the activity of nitrifying bacteria occurs. Therefore nitrifying bacteria are dependent on organotrophs to reduce cBOD.

Organotrophs are found in freshwater, potable water, wastewater, marine water, brackish water, and soil. They also are associated with plants and the fecal waste from humans and animals. Organotrophs enter sewer systems and activated sludge processes through domestic wastewater and I/I. Some organotrophs are disease-causing organisms.

Although it is difficult to enumerate and characterize the viable numbers of organotrophs within soil and fecal wastes, population sizes of these organisms are estimated as several billion per gram of soil or fecal waste. In the activated sludge process, organotrophs proliferate to relatively large numbers, approximately 60,000,000 per milliliter of bulk solution in the aeration tank and 20,000,000,000 per gram of MLVSS.

Organotrophs oxidize organic wastes through aerobic respiration or anaerobic respiration. Bacteria that can oxidize organic wastes through aerobic respiration only are strict or obligate aerobes; that is,

TABLE 8.2 Comparison of Carbon and Energy Substrates for Organotrophs and Nitrifying Bacteria

Carbon/Energy Substrate	Organotrophs	Nitrifying Bacteria
Carbon source	Organic wastes	CO_2 as alkalinity
Carbon source removal	Decrease cBOD	Decreases alkalinity/pH
Energy source	Organic wastes	NH_4^+ and NO_2^-
Energy source removal	Decreases cBOD	Decreases nBOD

the bacteria can only use free molecular oxygen. Bacteria that can oxidize organic wastes using free molecular oxygen, if it is available, or another molecule, such as nitrite ions or nitrate ions, if free molecular oxygen is not available are facultative anaerobes.

Although facultative anaerobes can use free molecular oxygen or another molecule to oxidize organic wastes, their preference always is for free molecular oxygen. There are two reasons for this preference. First, the organotrophs obtain a larger quantity of energy by oxidizing the organic wastes with free molecular oxygen rather than another molecule. Second, the organotrophs produce more offspring (greater reproduction) by oxidizing the organic wastes with free molecular oxygen rather than another molecule. Within the activated sludge process approximately 80% of the organotrophs are facultative anaerobes (Table 8.3).

The dominant genera of organotrophs in the activated sludge process are determined by wastewater composition. For example, proteinaceous wastes favor the proliferation of *Bacillus*, while carbohydrate wastes favor the proliferation of *Pseudomonas*.

TABLE 8.3 Common Genera of Organotrophs in the Activated Sludge Process

Genus	Strict Aerobes	Facultative Anaerobes
Achromobacter		×
Acinetobacter	×	
Actinomyces		×
Aerobacter		×
Arthrobacter	×	
Bacillus		×
Beggiatoa		×
Cornynebacterium		×
Enterobacter		×
Escherichia		×
Flavobacterium		×
Klebsiella		×
Micrococcus	×	
Nocardia	×	
Proteus		×
Pseudomonas		×
Sphaerotilus		×
Thiothrix	×	
Zoogloea		×

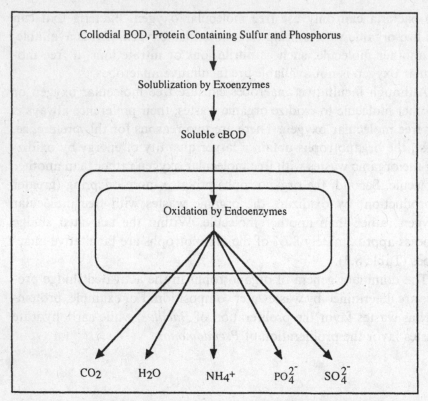

Figure 8.1 *Oxidation of soluble cBOD. Bacterial cells rapidly absorb soluble cBOD. Colloidal BOD such as proteins must be solublized through the production and release of bacterial exoenzymes. Bacterial cells rapidly absorb, once solublized, the resulting soluble cBOD. Inside the bacterial cells, endoenzymes oxidize the soluble cBOD into new cells, and several inorganic compounds including carbon dioxide (CO_2), water (H_2O), ammonium ions (NH_4^+), phosphate ions (PO_4^{2-}), and sulfate ions (SO_4^{2-}). Proteins that contain phosphate groups and thiol groups (–SH), serve as the compounds that yield phosphates and sulfates when oxidized.*

The dominant bacteria are selected by the compatibility of bacterial enzyme systems (ability to degrade substrates) and the substrates present in the wastewater. Enzyme systems are the "tools" or cellular machinery that bacteria use to degrade substrate. An enzyme system may degrade one substrate or a variety of substrates. However, no enzyme system can degrade all substrates, and no substrate can be degraded by all enzyme systems.

During the oxidation of cBOD, electrons are released from the substrate. The electrons are removed from the cell by free molecular

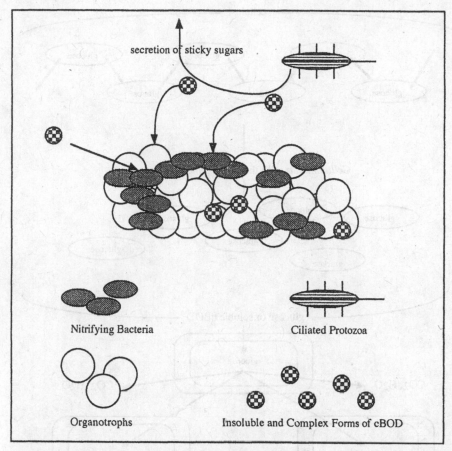

secretion of sticky sugars

Nitrifying Bacteria

Ciliated Protozoa

Organotrophs

Insoluble and Complex Forms of cBOD

Figure 8.2 *Removal of insoluble and complex forms of cBOD. Insoluble and complex forms of cBOD, namely particulate BOD and colloidal BOD, are adsorbed directly to the surface of floc particles if these forms of cBOD have compatible surface charge. These forms of cBOD are adsorbed indirectly to the surface of floc particles if their surface charge is made compatible through the coating action of secretions from ciliated protozoa and other higher life forms.*

oxygen, nitrite ions, or nitrate ions. If nitrite ions or nitrate ions are used to remove the electrons, molecular nitrogen (N_2) is produced. The production of molecular nitrogen from nitrite ions or nitrate ions during respiration is denitrification. For example, when nitrate ions are used to degrade methanol (CH_3OH), an organic substrate is degrade through anoxic respiration to produce carbon dioxide, water, molecular nitrogen, and new bacterial cells (Equation 8.1).

$$CH_3OH + NO_3^- - Bacillus \rightarrow 3CO_2 + N_2 + 9H^+ + 3OH^- \quad (8.1)$$

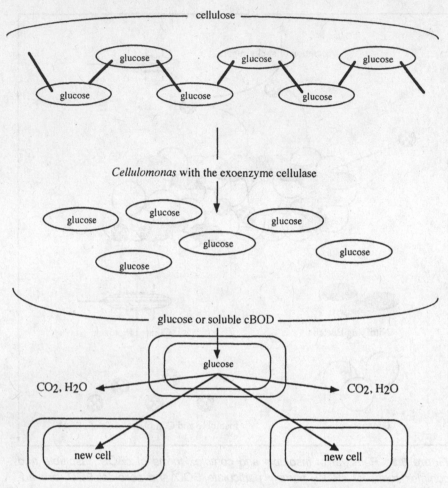

Figure 8.3 *Solublization and oxidation of cellulose. Cellulose is a starch having many glucose units bonded by special chemical bonds. Cellulose is insoluble in wastewater. Cellulose can be solublized and oxidized under appropriate operational conditions in the activated sludge process. If cellulose is adsorbed to the surface of the floc particles, and the bacterium Cellulomonas is present in the floc particle, then Cellulomonas produces and releases the exoenzyme cellulase. When cellulase comes in contact with cellulose, the special chemical bonds are broken, and individual glucose units are released. These glucose units quickly dissolve in the wastewater and are absorbed by Cellulomonas as well as other organotrophs. Once absorbed, cellulose is oxidized. The oxidation of cellulose results in the production of new organotrophs, carbon dioxide (CO_2), and water (H_2O).*

Molecular nitrogen is insoluble in wastewater, and when released in the secondary clarifier, it often becomes entrapped in the floc particles or solids. The entrapment of molecular nitrogen results in a loss of compaction of the solids and a loss of solids from the secondary clarifier.

When cBOD enters an activated sludge process, the organotrophs quickly absorb the soluble, simplistic forms of cBOD, such as acids and alcohols. This cBOD is then oxidized inside the bacterial cells (Figure 8.1). The insoluble and complex forms of cBOD are adsorbed to the slime on the surface of the bacterial cells (Figure 8.2). These forms of cBOD may be solublized and absorbed by the bacterial cells if sufficient retention time exists.

Exoenzymes are enzymes that are produced inside the cytoplasm of the bacterial cell and released through the cell membrane and cell wall to solublize the complex forms of cBOD (pBOD and coBOD) embedded in the slime. In order for the bacterial cells to produce and release exoenzymes, solublize complex forms of cBOD, and degrade the cBOD, approximately six hours of hydraulic retention time or HRT (Appendix II) are required in the aeration tank.

The solublization of cellulose (pBOD) is accomplished by the bacterium *Cellulomonas*. This bacterium produces the exoenzyme cellulase that is specific for the solublization of cellulose. When cellulose is solublized, soluble sugar (glucose) is produced (Figure 8.3). The sugar is easily absorbed by *Cellulomonas* as well as other organotrophs and is degraded.

Organotrophs also are capable of oxidizing some xenobiotics or synthetic compounds such as pesticides and chlorinated hydrocarbons. In addition to the oxidation of cBOD, some organotrophs such as *Achromobacter*, *Aerobacter*, *Bacillus*, *Escherichia*, *Flavobacterium*, and *Pseudomonas* initiate floc formation in the activated sludge process. The floc particles that develop eventually will contain large numbers of not only organotrophs but also nitrifying bacteria. Together these two groups of bacteria remove cBOD and nBOD from the activated sludge process.

9

The Wastewater Nitrogen Cycle

Because of the large number of valences or oxidation states of nitrogen, many nitrogenous compounds exist in the environment and the activated sludge process (Table 9.1). The majority of nitrogen found in the environment exists as molecular nitrogen. Approximately 80% of dry air by volume contains molecular nitrogen and represents an inexhaustible reservoir of this essential element.

Although constituting much less of the biomass than carbon or oxygen, nitrogen is an essential element for all living organisms. It is incorporated in cellular material that is used for growth, enzyme production, and genetic information. However, molecular nitrogen is formed by a triple bond that is very difficult for most organisms to break. Fortunately molecular nitrogen is made available to living organisms when the triple bond is broken by a group of unique bacteria and fixed or converted to ammonium ions.

The bacteria that convert molecular nitrogen to ammonium ions are nitrogen-fixing bacteria (Table 9.2). The bacteria may be free-living in the soil around the roots of plants or may grow in a symbiotic relationship in the nodules of the roots of green plants (Figure 9.1).

Nitrogen fixation, namely the conversion of molecular nitrogen to ammonium ions, is achieved by the enzyme nitrogenase that is found only in nitrogen-fixing bacteria (Equation 9.1). Prior to the widespread use of nitrogen fertilizers, nodule-growing plants or legumes provided soil nitrogen. Examples of legumes include alfalfa, clover, and soybeans.

TABLE 9.1 Nitrogenous Compounds and Their Oxidation States

Compound	Oxidation State	Chemical Name
NH_3	−3	Ammonia
NH_4^+	−3	Ammonium ion
NH_4OH	−3	Aqua ammonia
NH_4HCO_3	−3	Ammonium bicarbonate
N_2	0	Molecular nitrogen
NO_2^-	+3	Nitrite ion
NO_3^-	+5	Nitrate ion

$$3CH_2O + 2N_2 + 3H_2O + 4H^+ \rightarrow 3CO_2 + 4NH_4^+ \qquad (9.1)$$

The nodules that develop in the roots of the plant result from an "irritating" action caused by the nitrogen-fixing bacteria. The nodules connect directly to the vascular (circulatory) system of the plant. This connection enables a symbiotic or mutually advantageous relationship between the bacteria and the plant. The bacteria obtain photosynthetically produced energy directly from the plant, while the plant obtains ammonium ions. The ammonium ions are assimilated into plant material in the form of amino acids and proteins within tissues, roots, and seeds. The plant may be consumed and pass its ammonium ions along the food chain, or the plant may die and decompose. When the plant decomposes, it releases ammonium ions to the soil that can be used by other plants.

Some algae also can use molecular nitrogen for the production of amino acids and proteins. The algae take the molecular nitrogen from

TABLE 9.2 Nitrogen-Fixing Bacteria

Type of Nitrogen-Fixing Bacteria	Genus
Free-living bacteria	*Arthrobacter*
	Azobacter
	Azotobacter
	Azospirillum
	Clostridium
	Cyanobacterium
	Enterobacter
Symbiotic bacteria	*Frankia*
	Rhizobium

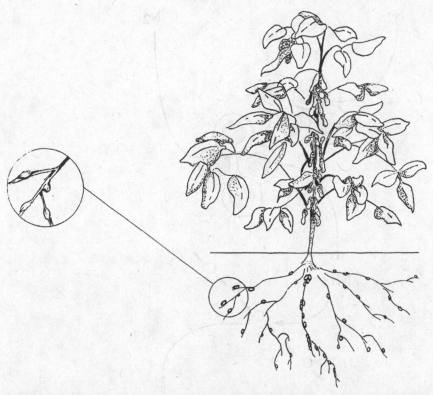

Figure 9.1 *Root nodules and nitrogen-fixing bacteria. Bacteria that fix molecular nitrogen (N_2) are found in the soil immediately around the roots of plants or in nodules growing on the surface of the roots of plants. These nitrogen-fixing bacteria convert molecular nitrogen to ammonium ions (NH_4^+). Nitrogen-fixing bacteria that live in the soil are free-living, while nitrogen-fixing bacteria that live in the nodules are symbiotic.*

the atmosphere and assimilate it with an organic molecule (Figure 9.2). Eventually these organic molecules with nitrogen incorporated into their structure are distributed throughout the food chain as the algae are consumed by higher life forms.

The movement of nitrogen and its changes in oxidation states from the atmosphere to living organisms to the activated sludge process and its return to the atmosphere is the wastewater nitrogen cycle (Figure 9.3). This cycle incorporates the following critical nitrogenous compounds: molecular nitrogen, amino acids, proteins, urea, ammonium ions, ammonia, nitrite ions, and nitrate ions. Amino acids and proteins are organic forms of nitrogen. Molecular nitrogen, ammonium ions, ammonia, nitrite ions, and nitrate ions are inorganic forms of nitrogen.

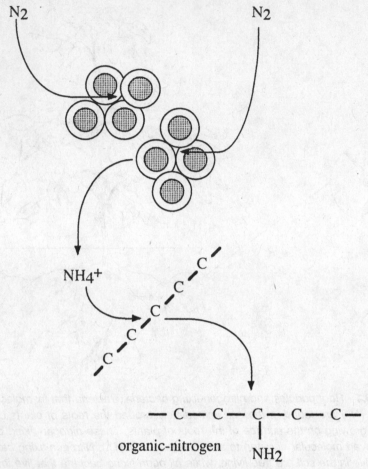

Figure 9.2 *Nitrogen-fixing by algae. A large number of algae, including those that grow in tetrad formation (colony of four) can fix molecular nitrogen. When algae fix molecular nitrogen, the nitrogen is converted to ammonium ions and then incorporated in an organic molecular to form amino acids and proteins.*

Municipal, activated sludge processes receive large quantities of organic-nitrogen wastes. Some of the organic-nitrogen wastes release ammonium ions in the sewer system as they are hydrolyzed and deaminated by organotrophs. Due to organotrophic activity within the sewer system, municipal, activated sludge processes have an influent ammonium ion concentration of usually 15 to 30 mg/l. Approximately 40% of the nitrogen in domestic wastewater is in the form of ammonium ions, and the rest of the nitrogen is primarily in the form of organic nitrogen.

The production of nitrite ions and nitrate ions within the sewer system is rare. Conditions in the sewer system are not favorable for the production or nitrification of these ions. Adverse conditions within the sewer system that prevent nitrification include a lack of adequate oxygen, a small nitrifying bacterial population, and a short retention time. However, relatively large quantities of nitrite ions and nitrate ions may be found in the sewer system, if they are discharged to the sewer system by industrial wastes containing these ions, such as steel mill wastewater.

The amino acids and proteins within plant tissues, roots, and seeds and livestock meats are discharged directly to the sewer system (garbage disposal waste and food processing wastewater) and indirectly to the sewer system (fecal waste). Many bacteria within the sewer system deaminate some amino acids and proteins (Equation 9.2). Deamination is achieved with the enzyme deaminase and results in the production of ammonium ions. The production of ammonium ions also is known as ammonification. Deamination of the amino acid phenylalanine is shown is Equation (9.3).

$$\text{Amino acid} \text{—} \text{Organotrophs} \rightarrow NH_4^+ + \text{Acid} \qquad (9.2)$$

$$\text{Phenylalanine} \text{—} \textit{Proteus} \rightarrow NH_4^+ + \text{Phenylpyruvic acid} \qquad (9.3)$$

Urea is an organic-nitrogen compound that is found in urine, fertilizers, and stockyard wastes. When hydrolyzed by the bacterial enzyme urease, ammonium ions are released (Equation 9.4). Urease is found in many organotrophs associated with fecal waste including *Citrobacter*. The hydrolysis of urea into ammonia and carbon dioxide by bacterial activity is very rapid. At the pH of sewer system ammonia is quickly converted to ammonium ions.

$$NH_2COHN_2 + H_2O \text{—} \textit{Citrobacter} \rightarrow 2NH_3 + CO_2 \qquad (9.4)$$

Amino acids and proteins that are not degraded in the sewer system may be degraded in the aeration tank. Degradation of amino acids or proteins in the aeration tank also results in the production of ammonium ions.

Ammonium ions within the activated sludge process have several fates (Figure 9.4). They may be used as a nutrient source for nitrogen by organotrophs and nitrifying bacteria. They may be air-stripped to the atmosphere as ammonia at high pH, and under appropriate operational conditions, *Nitrosomonas* may oxidize them to nitrite ions. If ammonium ions are not used as a nutrient source, air-stripped, or oxidized, they are discharged in the aeration tank effluent.

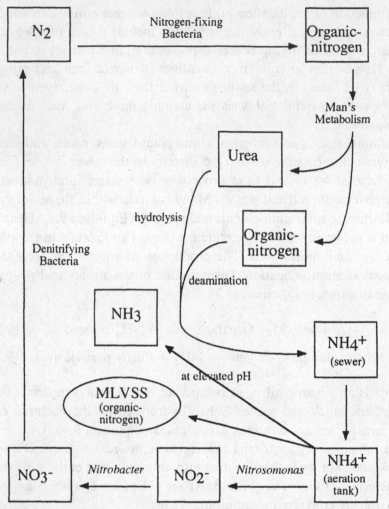

Figure 9.3 *Wastewater nitrogen cycle. In the wastewater nitrogen cycle, molecular nitrogen is removed from the atmosphere and returned to the atmosphere. The movement of nitrogen through the cycle involves a number of conversions (oxidation and reduction reactions) and a variety of bacteria. Molecular nitrogen (N_2) is removed from the atmosphere by nitrogen-fixing bacteria. These bacteria produce ammonium ions (NH_4^+) from molecular nitrogen. The ammonium ions produced by nitrogen-fixing bacteria are incorporated into amino acids and proteins within green, leafy plants. The plants (fruits, leaves, roots, seeds, and stems) are consumed by humans, and some of the nitrogen within the plants is released in human bodily wastes (urine and fecal material) into the sewer system in the form of urea, amino acids, and proteins. Some of these organic-nitrogen wastes undergo hydrolysis and deamination in the sewer system and release ammonium ions. Some of these organic wastes undergo deamination in the aeration tank and release ammonium ions. There are several fates for the ammonium ions within the aeration tank. If the pH of the tank increases to 9.4 or greater, some of the ammonium ions are converted to ammonia (NH_3) and are lost to the atmosphere.*

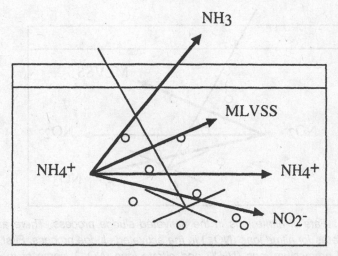

Figure 9.4 *Fate of ammonium ions in the activated sludge process. There are four significant fates for the ammonium ion (NH$_4^+$) in the activated sludge process. First, at a pH of 9.4 or higher, some of the ammonium ions in the aeration tank are converted to ammonia (NH$_3$). Ammonia escapes from the aeration tank in the form of a gas. Second, some of the ammonium ions are used as a nitrogen nutrient by the bacteria within the aeration tank for growth and reproduction, namely an increase in MLVSS or organic-nitrogen content of the process. Third, under favorable operational conditions, some of the ammonium ions are oxidized by* Nitrosomonas *to nitrite ions (NO$_2^-$). Fourth, the ammonium ions may leave the aeration tank and enter the secondary clarifier.*

Under cold temperature or a limiting process condition, nitrite ions may accumulate in the activated sludge process. Nitrite ions also have several fates in the activated sludge process (Figure 9.5). They may be chemically oxidized to nitrate ions if chlorine is being used to control the undesired growth of filamentous organisms. Nitrite ions may be biological oxidized by *Nitrobacter* under appropriate operational conditions. If ammonium ions and nitrate ions are not available in the aeration tank, nitrite ions may be used as a nutrient source

---

Some of the ammonium ions are used as a nitrogen nutrient for the growth of new bacteria or MLVSS (organic nitrogen). If operational conditions are favorable, Nitrosomonas *oxidizes some of the ammonium ions to nitrite ions (NO$_2^-$), and* Nitrobacter *oxidizes some of the nitrite ions to nitrate ions (NO$_3^-$). If an anoxic condition occurs within the treatment process, the nitrate ions are reduced to molecular nitrogen (N$_2$) by facultative anaerobic or denitrifying bacteria. Molecular nitrogen escapes from the treatment plant to the atmosphere through denitrification.*

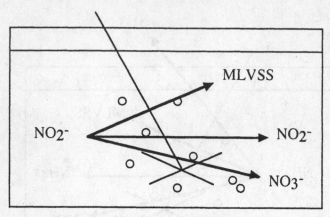

Figure 9.5 *Fate of nitrite ions in the activated sludge process. There are four significant fates for nitrite ions (NO_2^-) in the activated sludge process. First, in the absence of ammonium ions (NH_4^+) and nitrate ions (NO_3^-), some of the nitrite ions are used as a nitrogen nutrient by the bacteria within the aeration tank for growth and reproduction, meaning that an increase occurs in MLVSS or organic-nitrogen content of the process. Second, under favorable operational conditions, some of the nitrite ions are oxidized by Nitrobacter to nitrate ions. Third, if the MLVSS are being chlorinated to control undesired filamentous growth, some of the nitrite ions are chemical oxidized by chlorine to nitrate ions. Fourth, the nitrite ions may leave the aeration tank and enter the secondary clarifier. Here the nitrite ions may undergo denitrification if an anoxic condition develops in the secondary clarifier.*

for nitrogen by organotrophs. If nitrite ions are not oxidized or used as a nutrient source for nitrogen, they are discharged in the aeration tank effluent. In the secondary clarifier, nitrite ions may be denitrified.

Nitrate ions within the activated sludge process also have several fates (Figure 9.6). In the absence of ammonium ions in the aeration tank, nitrate ions may be used as a nutrient source for nitrogen. If nitrate ions are not used as a nutrient source for nitrogen, they are discharged in the aeration tank effluent. In the secondary clarifier, nitrate ions may be denitrified.

Nitrate ions are of central importance in the wastewater nitrogen cycle. They are the product of nitrification, the substrate of denitrification, and the source for the nitrogen nutrient when ammonium ions are not available. Nitrate ions are used as a nutrient source for nitrogen through a biological process known as nitrate assimilation. Nitrate ions are the most abundant, inorganic nitrogen source in most waters.

Nitrogenous compounds from industrial wastewater (Table 9.3) may release ammonium ions in the sewer system or activated sludge

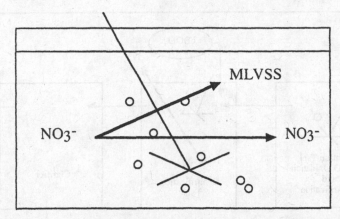

Figure 9.6 *Fate of nitrate ions in the activated sludge process. There are two significant fates for nitrate ions (NO_3^-) in the activated sludge process. First, in the absence of ammonium ions (NH_4^+), some of the nitrate ions are used as a nitrogen nutrient by the bacteria within the aeration tank for growth and reproduction, meaning that an increase occurs in MLVSS or organic-nitrogen content of the process. Second, the nitrate ions may leave the aeration tank and enter the secondary clarifier. Here the nitrate ions may undergo denitrification if an anoxic condition develops in the secondary clarifier.*

process depending on their structure and ease of bacterial degradation. Additional sources of nitrogenous compounds include polyelectrolytes and disinfectants, such as chloramines, added to potable water supplies.

Denitrification may occur in the sludge blanket of the secondary clarifier when an anoxic condition develops in the sludge blanket. Here facultative anaerobic bacteria use nitrite ions and nitrate ions to degrade soluble cBOD. This degradation is associated with the release of molecular nitrogen.

An activated sludge process that must satisfy a total nitrogen discharge requirement must nitrify and denitrify. However, denitrification is controlled. Denitrification occurs in a denitrification tank

TABLE 9.3   Examples of Nitrogenous Compounds Discharged by Industries

Nitrogenous Compounds	Source or Use
Acrylonitrile	Synthetic acrylic fibers
Benzoic acid	Fruit preservative
Dyes	Fabric and paper coloring
EDTA	Soap to remove metallic contaminates

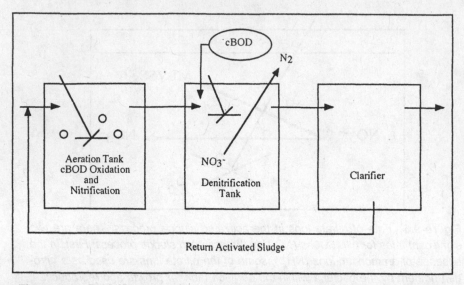

Figure 9.7 *Denitrification tank. A denitrification tank is a unit found immediately downstream or following the aeration tank in which nitrification has occurred. The nitrates produced in the aeration tank, as well as bacteria in the aeration tank, are discharged to the denitrification tank. In the denitrification tank, slow subsurface mixing is provided to rapidly remove residual dissolved oxygen and a carbon source or soluble cBOD such as methanol is added. Within the denitrification tank the facultative anaerobic or denitrifying bacteria discharged from the aeration tank use the nitrate ions to degrade the soluble cBOD added to the tank. When this occurs, molecular nitrogen is produced and released to the atmosphere. Denitrification within the tank takes approximately 30 to 60 minutes. The bacteria or solids within the denitrification tank are discharged to a clarifier where the settled solids are returned to the aeration tank.*

(Figure 9.7). MLVSS and nitrate ions produced in the aeration tank are added to the denitrification tank along with a source of soluble cBOD such as methanol. The denitrification tank is slowly mixed. Facultative anaerobic bacteria within the MLVSS degrade the soluble cBOD using nitrate ions. Nitrogen within the wastes entering the activated sludge process have been nitrified to nitrate ions, and the nitrogen within the nitrate ions is loss to the atmosphere as molecular nitrogen, not the receiving water, through the use of the denitrification tank.

Much of the proteinaceous waste associated with suspended solids in the wastewater is removed by sedimentation in primary clarifiers, and some is solublized to soluble cBOD in the aeration tank (Figure 9.8). Proteinaceous wastes from the primary clarifier and the acti-

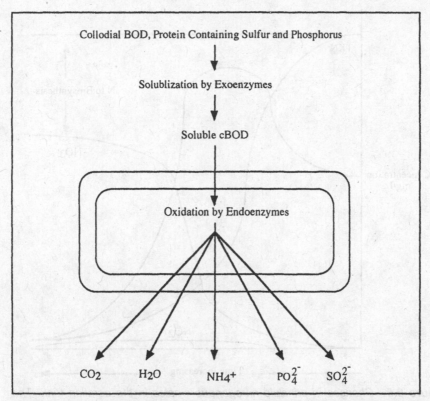

Figure 9.8 *Removal and degradation of proteinaceous waste. Proteins are colloidal in nature and must be solublized in order to enter bacterial cells and undergo degradation. Solublization of proteins is achieved through the use of exoenzymes. Once solublized, bacterial cells rapidly absorb soluble cBOD. Inside the bacterial cells, endoenzymes oxidize the soluble cBOD into new cells and several inorganic compounds including carbon dioxide (CO_2), water (H_2O), ammonium ions (NH_4^+), phosphate ions (PO_4^{2-}), and sulfate ions (SO_4^{2-}). Proteins that contain phosphate groups and thiol groups ($-SH$) serve as the compounds that yield phosphates and sulfates when oxidized.*

vated sludge process are degraded in aerobic digesters or anaerobic digesters.

Ammonium ions may be removed by mixing action or air stripping to the atmosphere as ammonia. However, the loss of ammonia through air-stripping is relatively small, that is, less than 10%.

An overview of the changes in quantities of nitrogenous wastes in the activated sludge process is illustrated in Figure 9.9. In a municipal, activated sludge process receiving no nitrite ions and no nitrate ions from an industrial source, the concentration of ammonium ions

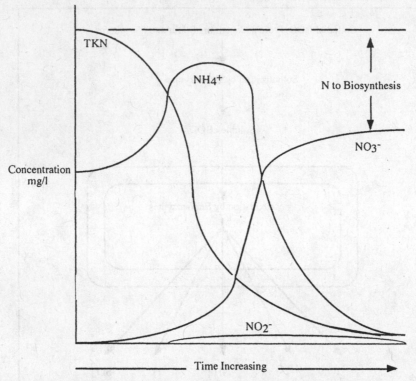

Figure 9.9 *Changes in principle nitrogenous wastes in the aeration tank. The principle nitrogenous wastes in an aeration tank are organic-nitrogen compounds or TKN, ammonium ions (NH_4^+), nitrite ions (NO_2^-), and nitrate ions (NO_3^-). As TKN and ammonium ions enter the aeration tank, TKN is deaminated to release ammonium ions, and bacteria remove ammonium ions as a nitrogen nutrient. However, TKN is deaminated more rapidly than ammonium ions are removed, and the amount of ammonium ions in the aeration tank initially increasing. Following this increase the amount of ammonium ions within the aeration tank begins to drop. The drop in the amount of ammonium ions is due to two factors. First, little TKN remains to replenish the amount of ammonium ions removed as a nitrogen nutrient. Second, some of the ammonium ions are oxidized to nitrite ions by Nitrosomonas. If no adverse operational condition develops in the aeration tank, nitrite ions do not accumulate and are oxidized to nitrate ions by Nitrobacter. With extended aeration or nitrification, the amount of ammonium ions within the aeration tanks continues to decrease, while the amount of nitrate ions continues to increase.*

and organic nitrogen is relatively high in the aeration tank. Although ammonium ions are removed as a nutrient source for nitrogen, its concentration increases as organic-nitrogen wastes are deaminated. As the quantity of organic-nitrogen wastes decreases, the quantity of ammonium ions increases.

When the organic-nitrogen wastes are no longer available to release ammonium ions, the quantity of ammonium ions decreases. The decrease in ammonium ions occurs as they are used as a nutrient source for nitrogen and oxidized to nitrite ions and nitrate ions. If nitrification is proceeding properly, no accumulation of nitrite ions occurs. Some of the nitrate ions may be removed as a nutrient source for nitrogen when ammonium ions are depleted. If denitrification occurs, the quantity of nitrate ions would be greatly reduced, perhaps eliminated.

10

Nitrogen Assimilation

In the activated sludge process nitrogen is utilized or assimilated by bacteria as a nutrient for several critical cellular functions (Table 10.1). Approximately 14% of the dry weight of most bacterial cells is nitrogen. Without the proper quantity and type of nitrogen, bacterial activity would be inhibited, and improper wastewater treatment would occur.

For any nutrient to enter a bacterial cell, the nutrient must be dissolved in the bulk solution surrounding the cell. Those nutrients that pass through the cell wall and cell membrane by following a concentration gradient of higher concentration outside the cell to lower concentration inside the cell are preferred (Figure 10.1). Preferred nutrients that move by a concentration gradient or diffusion are "readily available" nutrients; that is, they are immediately available for use and require no expenditure of energy to diffuse into the cell.

Nitrogen may enter bacterial cells as a component of a variety of organic and inorganic molecules or ions (Table 10.2). Most nitrogenous wastes, whether simple or complex in structure, can be enzymatically converted to ammonium ions. When ammonium ions are no longer available for use as the nitrogen nutrient, nitrate ions are used next. In order to use nitrate ions as the nitrogen nutrient, nitrate ions must be reduced to ammonium ions; that is, oxygen must be removed from the nitrate ions, and hydrogen added to the nitrogen to form ammonium ions inside the bacterial cell (Figure 10.2). Nitrogen always enters biosynthetic pathways (processes for producing cellular material) in an inorganic form and an oxidation state of -3.

TABLE 10.1 Cellular Functions of Nitrogen

Function	Example
Structural material for cellular growth	Pepidogylcans (cell wall components)
Structural material for enzymes	Cellulase, nitrogenase, urease
Transfer of genetic material	Nucleic acids

Ammonium ions are the preferred nutrient source of nitrogen. These ions are water soluble, enter the cell by diffusion, and are in the oxidation state of -3. Most organic sources of nitrogen, such as amino acids, contain nitrogen in the -3 oxidation state, but these sources must be degraded to yield ammonium ions. When ammo-

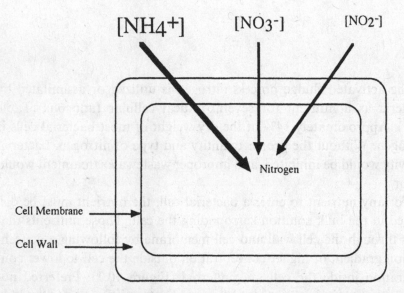

Figure 10.1 Preferred or readily available nutrients. Whenever bacterial cells use nitrogen for the synthesis of amino acids or proteins, nitrogen is incorporated into these organic-nitrogen compounds in the -3 oxidation state or valence. Ammonium ions (NH_4^+) contain nitrogen in the -3 oxidation state. Ammonium ions are highly water soluble, very motile in water, and diffuse or move quickly from an area of higher concentration to an area of lower concentration. Also the bacterial cells need not use any energy to bring the ammonium ions into the cell. Therefore, when nitrogen is needed and ammonium ions are available, ammonium ions are used before any other form of nitrogen. Nitrate ions (NO_3^-) and nitrite ions (NO_2^-) are used as a nitrogen nutrient after the ammonium ion concentration surrounding the cell as been exhausted. Nitrogen is in the $+5$ oxidation state in nitrate ions and $+3$ oxidation state in the nitrite ion. Although less time and energy are required by the bacteria to convert $+3$ nitrogen (nitrite ion) to -3 nitrogen than to covert $+5$ nitrogen (nitrate ion) to -3, the $+3$ nitrogen in the form of nitrite ions is highly unstable, meaning it quickly converts to nitrate ions through nitrification.

TABLE 10.2   Molecules and Ions Providing the Nitrogen Nutrient

Molecule or Ion	Organic Form	Inorganic Form	Readily-Available
Amino acids	×		
N_2		×	
NH_4^+		×	×
NO_2^-		×	×
NO_3^-		×	×

nium ions are present in the bulk solution, ammonium ions are used as the nutrient source for nitrogen, and organic-nitrogen compounds are not used.

Bacteria, as a group, use ammonium ions, and most bacteria can utilize nitrite ions and nitrate ions, but the use of nitrite ions and nitrate ions is somewhat restricted. The use of nitrite ions and nitrate ions as a nutrient source for nitrogen is referred to as assimilatory nitrate or nitrite reduction (Equation 10.1). In order to assimilate nitrogen from these ions, the nitrogen within the ions must be reduced

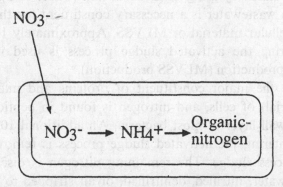

Figure 10.2 Use of nitrate ions as a nutrient source for nitrogen. Nitrate ions (NO_3^-) can be reduced biologically to molecular nitrogen (N_2) or ammonium ions (NH_4^+). When oxygen is stripped from nitrate ions to form molecular nitrogen as occurs during denitrification, nitrogen is not incorporated or assimilated in cellular material. This form of nitrate reduction is termed dissimilatory nitrate reduction. When ammonium ions (NH_4^+) are no longer available for use by bacteria as a nitrogen nutrient, nitrate ions are then used as the nitrogen nutrient. When nitrate ions are used as the nitrogen nutrient, oxygen is stripped from the nitrogen atom and hydrogen is added to the nitrogen atom. The addition of hydrogen results in the production of ammonium ions and a change in oxidation state for nitrogen from +5 in nitrate ions to −3 in ammonium ions. In this oxidation state or in the form of ammonium ions, nitrogen is assimilated into new cellular material such as amino acid and proteins (organic nitrogen). This form of nitrate reduction is termed "assimilatory nitrate reduction."

TABLE 10.3   Ammonium Ions, Nitrite Ions, and Nitrate Ions As Nutrient Sources for Nitrogen

Nutrient Source	Oxidation State of Nitrogen	Nutrient Available for Use due to the Oxidation State of Nitrogen
NH_4^+	−3	100%
NO_2^-	+3	70%
NO_3^-	+5	30%

from the +5 oxidation state (nitrate ions) and the +3 oxidation state (nitrite ions) to the −3 oxidation state of nitrogen within the ammonium ion.

$$NO_3^- \rightarrow NO_2^- \rightarrow NH_4^+ \qquad (10.1)$$

The reduction in oxidation state for each ion requires cellular energy. Therefore less bacterial growth or MLVSS production is achieved using nitrite ions and nitrate ions as compared to the bacterial growth obtained by using ammonium ions (Table 10.3).

Nitrogen in wastewater is a necessary constituent for the production of new cellular material or MLVSS. Approximately 14% of the nitrogen entering the activated sludge process is used in cellular growth and reproduction (MLVSS production).

Nitrogen is the major constituent of proteins and nucleic acids (genetic material) of cells, and nitrogen is found in peptidoglycans, the rigid cell wall layer of most bacteria. An additional 10% to 15% of nitrogen entering the activated sludge process is removed in the wasting of excess sludge. The remaining nitrogen is discharged to the receiving water, nitrified, denitrified, or air stripped to the atmosphere.

11

Forms of Nitrification

Nitrification may occur in an aeration tank after adequate, soluble cBOD degradation has occurred and several operational conditions are satisfactory for nitrification to occur. These conditions include the presence of an adequate amount of dissolved oxygen, substrate (ammonium ions or nitrite ions) for nitrifying bacteria, adequate retention time of the wastewater in the aeration tank, and a healthy and active population of nitrifying bacteria. Nitrification more easily occurs in the presence of high MLVSS and high temperatures.

Under favorable operating conditions an activated sludge process may nitrify, because the influent total nitrogen is considerably greater than the MLVSS requirement for the nutrient nitrogen. There are several forms of nitrification that are identified by the amount of ammonium ions, nitrite ions, and nitrate ions produced in the aeration tank or found in the aeration tank or mixed liquor effluent (Table 11.1). Testing for these three nitrogenous ions may be performed rapidly. Unless industries discharge wastes that contain nitrite ions or nitrate ions to the sewer system, testing for these two nitrogenous ions in the influent of the aeration tank is seldom performed.

COMPLETE NITRIFICATION

If the ammonium ion concentration and the nitrite ion concentration in the mixed liquor effluent are <1 mg/l each and the nitrate ion

TABLE 11.1 Forms of Nitrification

| Form of Nitrification | Aeration Tank or Mixed Liquor Effluent | | |
	NH_4^+, mg/l	NO_2^-, mg/l	NO_3^-, mg/l
Complete	<1	<1	As great as possible
Incomplete #1	<1	As great as possible	<1
Incomplete #2	>1	<1	>1
Incomplete #3	<1	>1	>1
Incomplete #4	>1	>1	>1

concentration is as great as possible, nitrification has occurred and is considered to be complete. Since most municipal, activated sludge processes usually receive 15 to 30 mg/l of influent ammonium ions and more ammonium ions are released in the aeration tank, the expected nitrate ion concentration in the mixed liquor effluent should be >10 mg/l. In industrial, activated sludge processes that experience complete nitrification, the nitrate ion concentration in the mixed liquor effluent may be <10 mg/l, depending on the quantity and types of nitrogenous compounds in the waste stream.

If complete nitrification is not required of the activated sludge process and the activated sludge process uses chlorine for disinfection, it may be beneficial to leave some concentration of ammonium ions in the mixed liquor effluent. It is less costly to have a small concentration of ammonium ions to combine with the chlorine being added to the wastewater for disinfection than to disinfect without an ammonium ion residual. Also, without ammonium ions in the effluent, more chlorine will be needed to maintain the chlorine residual required by a discharge permit.

INCOMPLETE NITRIFICATION

Often an activated sludge process may experience incomplete (partial) nitrification resulting in operational upsets, increased operational costs, and permit violations. Incomplete nitrification may be seasonal; that is, it may be related to changes in temperature or may be related to a specific operational condition, such as slug discharge of soluble cBOD or temporary low dissolved oxygen level. Because there are two different nitrifying bacteria, *Nitrosomonas* and *Nitrobacter*, and two different biochemical reactions involved in nitrification, there are four forms of incomplete nitrification.

TABLE 11.2 Factors Responsible for Incomplete Nitrification

Form of Incomplete Nitrification	Operational Factor Responsible for Incomplete Nitrification
Incomplete #1	Theoretical (not likely to occur)
Incomplete #2	Limiting factor
Incomplete #3	Usually cold temperature
Incomplete #4	Limiting factor or cold temperature

By sampling and testing the mixed liquor effluent filtrate for the concentrations of ammonium ions, nitrite ions, and nitrate ions, complete nitrification or the form of incomplete nitrification can be determined. By identifying the form of incomplete nitrification occurring in the aeration tank, the operational factor responsible for its occurrence also can be identified and corrected (Tables 11.2 and 11.3). For example, an activated sludge process that is required to completely nitrify and is experiencing difficulty may be able to identify the operational factor responsible for incomplete nitrification. Once the factor is corrected, the activated sludge process can return to complete nitrification.

To determine the form of nitrification occurring in the aeration tank, a sample of mixed liquor effluent should be collected and filtered. The filtrate from the mixed liquor should be tested for the concentrations of ammonium ions, nitrite ions, and nitrate ions. Testing of the filtrate provides for a quick, simplistic, inexpensive, and reliable means of determining the form of nitrification in the aeration tank. Ammonium ion concentration can be analyzed using the probe method, while nitrite ion concentration and nitrate ion concentration can be analyzed using colorimetric or spectrophotometric methods.

TABLE 11.3 Limiting Factors Responsible for Incomplete Nitrification

Cold temperature
Deficiencies of key nutrients
High influent ammonium ion concentration
Inhibition/toxicity
pH swing or extreme pH
Short retention time in the aeration tank
Slug discharge of soluble cBOD
Temporary low dissolved oxygen level

As the bacteria in the mixed liquor suspended solids are removed, the filtrate will not contain ammonium ions used as the nitrogen nutrient that may have been released by the bacteria during testing. Therefore these ammonium ions are not measured. Since the mixed liquor effluent is used, no significant loss of nitrite ions or nitrate ions would have occurred through denitrification. If a sample of clarifier effluent had been used, loss of these two ions through denitrification may have occurred.

The wastewater sample selected for testing should not be taken after chlorination or dechlorination. Chlorine oxidizes nitrite ions to nitrate ions, while dechlorination reduces nitrate ions to nitrite ions. These samples do not provide for a proper measurement of the amounts of nitrite ions and nitrate ions that were produced biologically in the aeration tank. Chemicals commonly used for dechlorination are sulfur dioxide (SO_2) and sodium bisulfite ($NaHSO_3$).

Although a rate of nitrification can be determined easily in a laboratory where only nitrifying bacteria are involved in a reaction and all operational conditions are controlled, a rate of nitrification can be difficult to determined in an activated sludge process. In this case there exists a mixed culture of bacteria, organotrophs and nitrifying bacteria, and operational conditions change often. Due to the presence of a mixed culture, changing operational conditions, and the numerous fates of the ammonium ions, nitrite ions, and nitrate ions in the aeration tank, it is preferred to express nitrification in an activated sludge process as a form rather than a rate.

The release of ammonium ions in the aeration tank further complicates a rate determination for nitrification. Ammonium ions are produced in the aeration tank through deamination of organic-nitrogen compounds. If the production of ammonium ions is significant, the ammonium ion concentration leaving the aeration tank may be higher than the ammonium ion concentration entering the aeration tank.

Also cationic (polyacrylamide) polymers that enter an aeration tank are degraded. Cationic polymers are commonly used at activated sludge processes for sludge thickening, solids capturing, and sludge dewatering. If cationic polymers are overdosed or misapplied, the excess polymer may eventually end up in the aeration tank and undergo degradation. The degradation of cationic polymers results in the production of ammonium ions (Figures 11.1 and 11.2).

Nitrification is the oxidation of ammonium ions and nitrite ions. An activated sludge process nitrifies if the production of nitrite ions or nitrate ions occurs in the aeration tank. The production of these

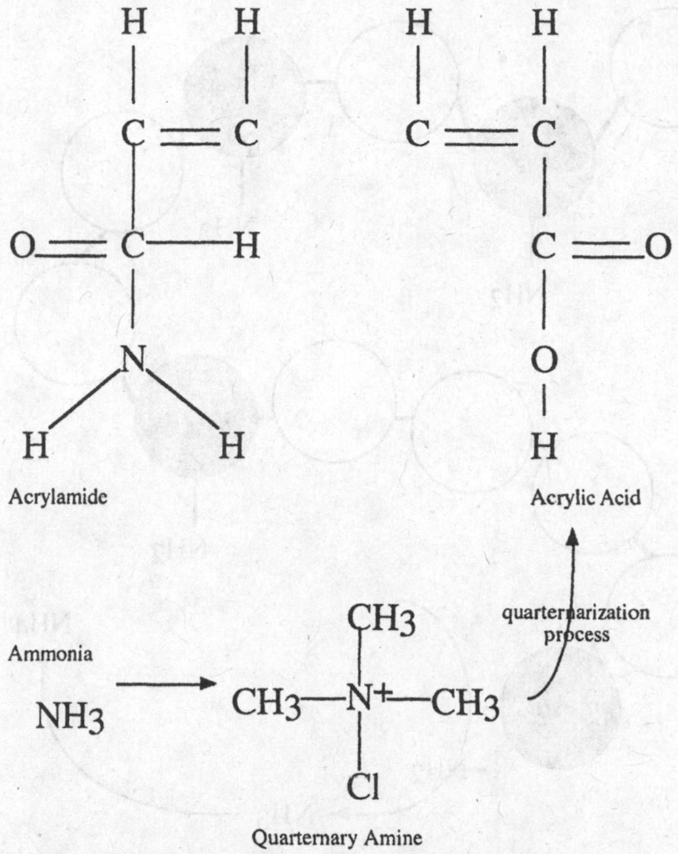

Figure 11.1 *Chemical components of polyacrylamide polymers. The two basic, chemical compounds used in the production of polyacrylamide polymers are acrylamide and acrylic acid. When these two compounds are joined together to form a very long polymer chain, the resulting polymer is a negatively charged or anionic polymer. To produce a positively charge polymer or cationic polymer, the carboxyl group (–COOH) on the acrylic acid undergoes a quarternarization process using ammonia. This process converts the negative charge of the carboxyl group to a positive charge.*

ions can be demonstrated by testing for their presence in the aeration tank effluent. The reduction in ammonium ion concentration across an aeration tank does not demonstrate nitrification. Nitrification is not the reduction of ammonium ions.

An activated sludge process that is required to nitrify usually tests the ammonium ion concentration in the final effluent to determine compliance with its permit requirement. However, the activated sludge process probably does not test for ammonium ion concentration, nitrite ion concentration, and nitrate ion concentration leaving

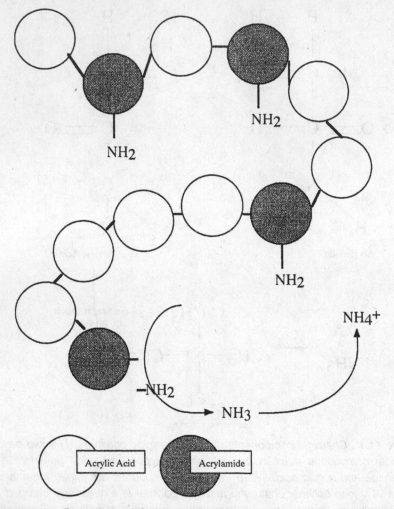

Figure 11.2 *Degradation of polyacrylamide polymers and ammonium ion release. When cationic polyacrylamide polymers are degraded in an aeration tank, the amino groups on the polymer are released. The released amino groups are converted quickly to ammonium ions in the aeration tank.*

the aeration tank effluent. If testing for these nitrogenous ions were preformed on a routine basis, the test results could be used to

- demonstrate nitrification,
- determine the form of nitrification,
- determine the limiting factor for incomplete nitrification,
- determine the corrective factor for incomplete nitrification, and
- provide for cost-effective operation.

12

Indicators of Nitrification

Activated sludge processes nitrify if nitrite ions or nitrate ions are produced in the aeration tank. Despite the various mechanisms of ammonium ion loss and production in an aeration tank (Table 12.1), a decrease in ammonium ion concentration across an aeration tank does not indicate nitrification. Nitrification is demonstrated by the production of nitrite ions or nitrate ions. However, if testing for nitrite ions or nitrate ions in the mixed liquor effluent is not performed, nitrification may be suspected by the presence of biological, chemical, and physical indicators (Table 12.2).

Biological indicators of nitrification include (1) the growth of algae and duckweed in secondary clarifiers, (2) a significant increase in mixed liquor oxygen demand, and (3) a significant decrease in mixed liquor dissolved oxygen level. Algae and duckweed obtain their nitrogen nutrient from nitrate ions. Therefore their presence in secondary clarifiers is an indicator of nitrate production or nitrification in the aeration tank.

Duckweed is the smallest and the simplistic of the flowering plants (Figure 12.1). Duckweed reproduces rapidly and floats on the water surface. It is used to control alga growth on oxidation ponds and remove nitrate ions in effluent lagoons before the effluent is discharged to receiving water. There are three genera of duckweed that are found in activated sludge process. These genera are *Lemna*, *Spirodella*, and *Wolffia*.

When nitrification occurs in an aeration tank, large quantities of

87

TABLE 12.1   Ammonium Ion Loss and Production in an Aeration Tank

Mechanism for Ammonium Ion Loss or Production	Ammonium Ion Loss	Ammonium Ion Production
Air-stripping of ammonia at high pH	×	
Deamination of amino acids and proteins		×
Degradation of cationic polymers		×
Nitrified to nitrite ions	×	
Nutrient source for nitrogen for bacteria	×	

dissolved oxygen are consumed by nitrifying bacteria as they oxidize ammonium ions and nitrite ions. Therefore, when nitrification occurs, a significant biomass demand for dissolved oxygen occurs, and the dissolved oxygen level within the aeration tank decreases.

Chemical indicators of nitrification include (1) an increase in chlorine demand for disinfection of the secondary effluent or control of undesired filamentous organisms within the mixed liquor or return activated sludge (RAS), (2) a decrease in mixed liquor alkalinity/pH, and (3) an increase in secondary clarifier alkalinity/pH. If incomplete nitrification occurs and nitrite ions accumulate, the nitrite ions react quickly with the chlorine, resulting in poor coliform kill in the secondary effluent and poor filamentous organism control in the mixed liquor or RAS. The production of nitrite ions within the aeration tank results in destruction of alkalinity and a drop in pH. Denitrification in the secondary clarifier (as a result of nitrification in an aeration tank) returns alkalinity to the wastewater. The return of alkalinity results in an increase in alkalinity/pH.

TABLE 12.2   Indicators of Nitrification

Indicator	Condition
Biological	Growth of algae in the clarifier
	Growth of duckweed in the clarifier
	Decreased mixed liquor dissolved oxygen level
	Increased mixed liquor dissolved oxygen demand
Chemical	Increased chlorine demand for control of filamentous organisms
	Increased chlorine demand for final effluent disinfection
	Decreased mixed liquor alkalinity/pH
	Increased secondary clarifier alkalinity/pH
Physical	Denitrification: Molecular nitrogen in secondary clarifier
	Denitrification: Sludge bulking in secondary clarifier

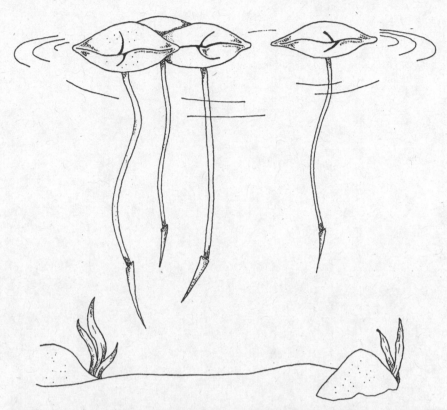

Figure 12.1 *Duckweed. Duckweed is the smallest flowering plant. The plant produces a white flower. Duckweed floats on the water with it "root" system suspended beneath its leaves. The nitrogen nutrient for duckweed is the nitrate ion (NO_3^-).*

Physical indicators of nitrification include (1) sludge bulking or clumping in the secondary clarifier and (2) the presence of molecular nitrogen rising to the surface of the secondary clarifier. Sludge bulking or large clumps of dark solids rising to the surface of the secondary clarifier is the result of denitrification. The solids rise to the surface, because large numbers of insoluble molecular nitrogen are trapped in the solids. The molecular nitrogen is produced through the denitrification of nitrite ions or nitrate ions, which are produced through nitrification in the aeration tank.

If any of the biological, chemical, or physical indicators of nitrification occur in an activated sludge process that is not required to nitrify, then nitrification should be suspected. To determine the form of nitrification occurring, mixed liquor effluent testing for the concentrations of ammonium ions, nitrite ions, and nitrate ions should be performed.

13

Nitrite Ion Accumulation

The production and accumulation of nitrite ions in an activated sludge process can occur during some forms of incomplete nitrification. The accumulation of nitrite ions is due to partial inhibition of enzymatic activity within nitrifying bacteria. This inhibition prevents the rapid oxidation of nitrite ions to nitrate ions. Operational factors responsible for this inhibition include cold temperature, deficiencies in key nutrients, high influent ammonium ion concentration, inhibitory and toxic wastes, pH changes, short retention time in the aeration tank, slug discharge of soluble cBOD, and temporary low dissolved oxygen level.

Nitrite ion accumulation in an activated sludge process may be seasonal in occurrence due to a change in temperature and MLVSS concentration. Nitrite ion accumulation usually occurs during late winter or early spring in an activated sludge process with an ammonia discharge limit. Nitrite ion accumulation usually occurs during late spring or early summer and late winter in an activated sludge process without an ammonia discharge limit.

Nitrite ions in relatively low concentrations may be toxic to not only many aquatic organism but also many wastewater organisms, including floc-forming bacteria. The critical concern with the accumulation of nitrite ions is the increased chlorine demand that occurs as a result of the nitrite ion accumulation. This increase in chlorine demand is known as the "chlorine sponge" or "nitrite kick." Nitrite ions interfere with the chlorination of the final effluent and undesired filamentous organisms. The interference with chlorination of the final effluent may result in a violation of the discharge permit for coliform bacteria.

Chlorine is a strong oxidizer. If *Nitrobacter* does not oxidize nitrite ions to nitrate ions, chlorine will oxidize the ions. Therefore an activated sludge process that produces and accumulates nitrite ions, meaning it experiences incomplete nitrification, will have variations in chlorine residuals and possible problems with the use of chlorine to achieve coliform kills and filamentous organism control.

When chlorine is added to the effluent for disinfection, hypochlorous acid (HOCl) is formed (Equation 13.1). Hypochlorous acid is the "free available chlorine" and the effective bactericide or killing agent of coliform bacteria and disease-causing organisms. However, the hypochlorous acid is a relatively weak acid and dissociates under the dilute aqueous solution that exists in the final effluent. When hypochlorous acid dissociates, the acid releases a hydrogen ion and a hypochlorite ion (Equation 13.2).

$$Cl_2 + H_2O \rightarrow HCl + HOCl \tag{13.1}$$

$$HOCl \leftrightarrow H^+ + OCl^- \tag{13.2}$$

Below pH 7.5 the hypochlorous acid predominates over the hypochlorite ion (OCl^-) (Figure 13.1). However, before the hypochlorous acid can effectively kill coliform bacteria, the hypochlorite ion reacts quickly with the nitrite ion. In this reaction the nitrite ion is oxidized to the nitrate ion, and the hypochlorite ion is reduced to the chloride ion (Cl^-) (Equation 13.3). As more and more hypochlorite ions react with the nitrite ions, more hypochlorous acid dissociates. With more and more dissociation of hypochlorous acid, less destruction of coliform bacteria occurs. The hypochlorite ion and the chloride ion are poor bactericides.

$$OCl^- + NO_2^- \rightarrow NO_3^- + Cl^- \tag{13.3}$$

The production and accumulation of nitrite ions not only adversely affect permit compliance for a coliform requirement but also interferes with process control. The "chlorine sponge" often is responsible for the inability of chlorine to effectively control undesired growth of filamentous organisms.

If the accumulation of nitrite ions is not excessive, then the

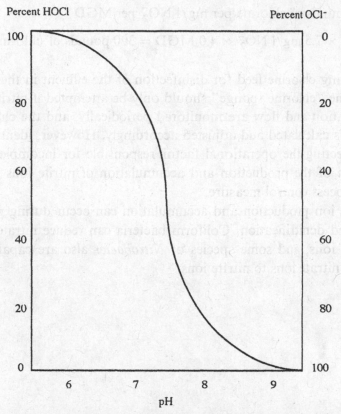

Figure 13.1 *Relative quantities of HOCl and OCl⁻ as affected by pH. The relative amount of HOCl and OCl⁻ within the chlorine contact tank, or the mixed liquor, is influenced strongly by the pH of the wastewater. At pH values less than 7.5, HOCl dominates, while at pH values greater than 7.5, OCl⁻ dominates.*

amount of chlorine needed to satisfy the "chlorine sponge" can be calculated. The amount of chlorine needed is approximately 13 pounds per milligram per liter (mg/l) of nitrite ion produce per million gallons per day of flow.

Chlorine Sponge

Calculate the amount of chlorine consumed in the chlorine contact tank of an activated sludge process by the presence of 7.5 mg/l of NO_2^- in the secondary effluent? The activated sludge process has a flow of 4.0 MGD.

The amount of chlorine consumed by the "chlorine sponge" is 13 pounds per milligram per liter (mg/l) of NO_2^- per MGD. Therefore the amount of chlorine consumed by the "sponge" is

13 pounds of chlorine per mg/l NO_2^- per MGD

$$\times\ 7.5\ mg/l\ NO_2^-\ \times 4.0\ MGD = 390\ pounds\ of\ chlorine$$

Adjusting chlorine feed for disinfection of the effluent in the presence of the "chlorine sponge" should only be attempted if nitrite ion concentration and flow are monitored periodically, and the chlorine demand is calculated and adjusted accordingly. However, identifying and correcting the operational factor responsible for incomplete nitrification of the production and accumulation of nitrite ions is the better process control measure.

Nitrite ion production and accumulation can occur during nitrification and denitrification. Coliform bacteria can reduce nitrate ions to nitrite ions, and some species of *Nitrobacter* also are capable of reducing nitrate ions to nitrite ions.

14

BOD

There are several, bacterial substrates that enter an activated sludge process. These substrates provide carbon and energy for bacterial activity, growth, and reproduction. The strength of each substrate is measured as milligrams per liter of biochemical oxygen demand (Table 14.1). The types of biochemical oxygen demand or BOD that are found in an activated sludge process include total (tBOD), particulate (pBOD), soluble (sBOD), colloidal (coBOD), carbonaceous (cBOD), and nitrogenous (nBOD) (Figure 14.1).

BOD not only provides energy for organotrophic bacteria and nitrifying bacteria but also provides energy for the higher life forms in the activated sludge process including ciliated protozoa, rotifers, and free-living nematodes (Figure 14.2). The higher life forms obtain carbon and energy when the ciliated protozoa consume bacteria, and the

**TABLE 14.1 Types of Biochemical Oxygen
Demand**

Type	Acronym
Total	tBOD
Particulate	pBOD
Soluble	sBOD
Colloidal	coBOD
Carbonaceous	cBOD
Nitrogenous	nBOD

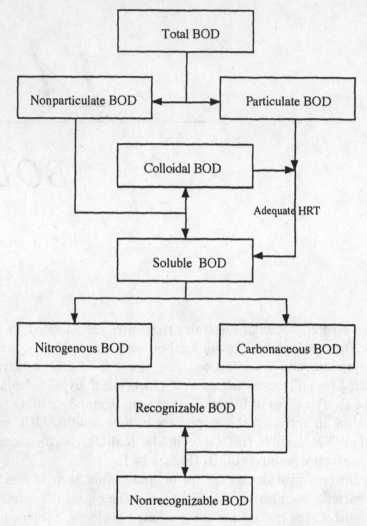

Figure 14.1 *Types of BOD. There are several types of BOD that enter an activated sludge process. Total BOD is the sum of all types of BOD found in the influent to the activated sludge process. Forms of particulate BOD are the solids such as cellulose that can be degraded. Much particulate BOD is removed in the primary clarifier, and unless adequate HRT is provided in the aeration tank, particulate BOD is adsorbed to the surface of floc particles in the aeration tank and is not degraded. Forms of nonparticulate BOD are soluble BOD such as ammonia ions and sugars and colloids such as proteins and lipids. Soluble BOD passes through the primary clarifier into the aeration tank. Colloidal BOD that is adsorbed to solids that settle in the primary clarifier is removed in the primary clarifier. Colloidal BOD that enters the aeration tank, like particulate BOD, is not degraded unless adequate HRT is provided. Colloidal BOD is adsorbed to the surface of floc particles in the aeration tank. There are two forms of soluble BOD: nitrogenous and carbonaceous. Nitrogenous BOD consists of ammonium ions and nitrite ions. These two ions can be oxidized under appropriate conditions in the aeration*

rotifers and free-living nematodes consume bacteria and protozoa. Collectively the bacteria and the higher life forms make up the food web in the activated sludge process (Figure 14.3).

Protozoa that are commonly found in activated sludge processes that nitrify are *Epistylis* and *Vorticella*. Besides rotifers and free-living nematodes, other multicellular, microscopic organisms found in activated sludge process that nitrify include bristleworms, flatworms, and waterbears (Figure 14.4). In order for these higher life forms to be present and active in an activated sludge process, the process must be stable. Activated sludge processes that nitrify are stable; that is, they contain a relatively high concentration of MLVSS and dissolved oxygen, low effluent cBOD, and no inhibition or toxicity.

The higher life forms are strict aerobes and are free-living soil and water organisms that enter the activated sludge process through I/I. The organisms are sensitive to changes in dissolved oxygen concentration and the presence of inhibitory or toxic wastes. Because nitrifying bacteria also are strict aerobes and are sensitive to changes in dissolved oxygen concentration and the presence of inhibitory or toxic wastes, the presence of higher life forms is expected during nitrification, and their absence or inactivity is expected when nitrification is lost.

Substrates immediately available to the bacteria and the food web consist of soluble cBOD. This type of BOD passes quickly through the cell wall and cell membrane of bacteria and is easily degraded. Forms of soluble cBOD that enter bacterial cells include simple acids, alcohols, and sugars.

Particulate BOD is made available to bacteria and the food web only after it has been solublized into simple molecules that can enter the bacterial cell. Solublization of pBOD occurs if adequate hydraulic retention time (HRT) is provided in the aeration tank and the

---

tank. Carbonaceous BOD consists of two forms: recognizable and nonrecognizable. Recognizable forms of carbonaceous BOD are simplistic one-, two-, three-, or four-carbon unit acids and alcohols. These forms of BOD can enter the cells of nitrifying bacteria and inhibit the enzymatic systems of the bacteria that oxidize ammonium ions and nitrite ions. Nonrecognizable forms of BOD are numerous and include sugars, amino acids, and long-chain acids and alcohols. These forms of BOD cannot enter the cells of nitrifying bacteria and cause inhibition. Recognizable and nonrecognizable forms of BOD can be oxidized under appropriate conditions in the aeration tank.

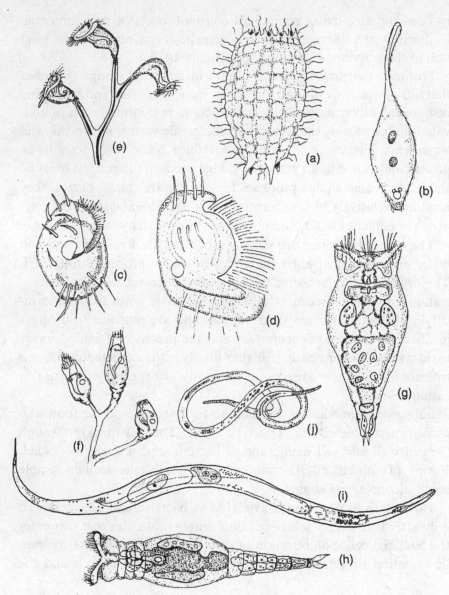

Figure 14.2 *Ciliated protozoa, rotifers, and free-living nematodes. Examples of ciliated protozoa that commonly are found in an activated sludge process include the free-swimming ciliates* Coleps *(a) and* Litonotus *(b), the crawling ciliates* Aspidisca *(c) and* Euplotes *(d), and the stalked ciliates* Carchesium *(e) and* Opercularia *(f). Free-swimming ciliates have cilia on all surfaces of the body, while crawling ciliates only have cilia on the ventral or "belly" surface, and stalked ciliates have cilia only around the mouth opening. Rotifers and free-living nematodes (worms) are found less frequently and in small numbers in an activated sludge process than ciliated protozoa. Examples of rotifers that commonly are found in an activated sludge process include* Daphnia *(g) and* Philodina *(h). Free-living nematodes are found in the activated sludge process in an elongated form (i) and coiled form (j).*

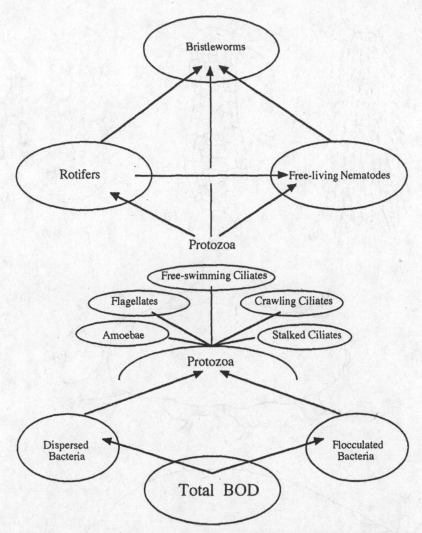

Figure 14.3 *The food web. Within the activated sludge process or aeration tank there is a food web in which energy and carbon are transferred "upward" from one organism to another organism. The basis for the food web is the BOD that enters the aeration tank. The BOD contains carbon and energy. The energy is found in the form of chemical bonds. As bacteria break the chemical bonds, the bacteria obtain carbon and energy for life, and they reproduce. Protozoa consume the bacteria. The carbon and chemical energy within the bacteria are transferred to the protozoa. There are five groups of protozoa that eat bacteria. These groups are the amoebae, flagellates, free-swimming ciliates, crawling ciliates, and stalked ciliates. Higher life forms (rotifers and free-living nematodes) consume the protozoa. The carbon and chemical energy within the protozoa is transferred to the rotifers and free-living nematodes. The carbon and chemical energy within the rotifers and free-living nematodes is again transferred to a higher life form— the bristleworms. The transfer of carbon and energy throughout a community of organisms is known as the food web.*

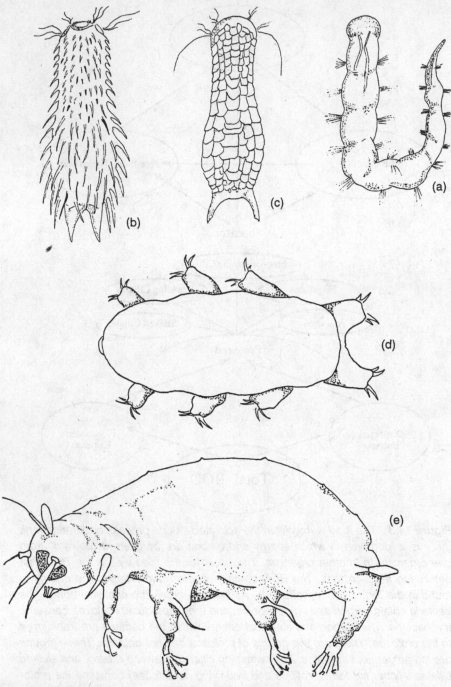

Figure 14.4 *Bristleworm, flatworms, and waterbears. The bristleworm (a) in the activated sludge process is segmented and possesses hairy structures or bristles that enable the worm to move through the soil. Flatworms appear in several interesting forms (b and c). The worm is "flat" only on the ventral or "belly" surface, while the curved dorsal or "back" surface may possess "spines" (b) or scalelike*

TABLE 14.2   Examples of Recognizable Soluble cBOD

cBOD	Chemical Formula	Number of Carbon Units
Methanol	CH_3OH	1
Methylamine	CH_2NH_2	1
Ethanol	CH_3CH_2OH	2
n-Propanol	$CH_3CH_2CH_2OH$	3
i-Propanol	$(CH_3)_2CHOH$	3
n-Butanol	$CH_3(CH_2)_2CH_2OH$	4
t-Butanol	$(CH_3)_3COH$	4
Ethyl acetate	$CH_3CO_2C_2H_5$	4
Aminoethanol	$CH_3NH_2CH_2OH$	2

necessary enzymes for solublizing pBOD are produced. For example, cellulose cannot enter bacterial cells unless the cellulase enzyme is produced and solublizes it.

Colloidal BOD is made available to bacteria and the food web only after it too has been solublized into simple molecules that can enter the bacterial cell. Colloidal BOD is solublized in similar fashion as pBOD. Examples of coBOD are proteins and lipids. When proteins are solublized, amino acids are produced. When lipids are solublized, fatty acids are produced. Simple amino acids and simple fatty acids can enter bacterial cells.

Particulate BOD and colloidal BOD that are not solublized in the aeration tank are wasted from the tank. Wasting of pBOD and coBOD occurs when each form of BOD is adsorbed to floc particles, and the floc particles are wasted from the activated sludge process to a digester. It is in the digester where large quantities of pBOD and coBOD are degraded.

Soluble cBOD consists of two types, recognizable and nonrecognizable. The difference between these two types of cBOD is based on

◄───

plates (c). Many flatworms have a forked tail. Waterbears are an odd-looking organism. From the dorsal (d) or lateral (e) appearance of the waterbear, four appendages or "legs" can be observed. Clawlike structures may be found at the end of each appendage. Bristleworms, flatworms, and waterbears are soil and water organisms that enter activated sludge processes through inflow and infiltration. They are strict aerobes and are "sensitive" to the lowest concentrations of inhibitory or toxic wastes. Therefore these higher life forms are found in large, active numbers under stable operating conditions.

their ability to inhibit nitrification. Recognizable, soluble forms of cBOD are simplistic molecules that are able to enter the cells of nitrifying bacteria and inactivate their enzyme systems (Table 14.2). This form of cBOD must be degraded significantly or completely by organotrophs in the aeration tank in order for nitrify bacteria to oxidize ammonium ions and nitrite ions. Nonrecognizable, soluble forms of cBOD are large, complex molecules such as proteins and starches that are not able to enter the cells of nitrifying bacteria.

15

Dissolved Oxygen

Although all bacteria require water, only a few actually require dissolved oxygen. The lack of dependence of bacteria on dissolved oxygen is due in part to the fact that oxygen is not highly soluble in water. Therefore most bacteria are capable of using a molecule other than dissolved oxygen for the degradation of substrate, that is, respiration.

Dissolved oxygen is the free or chemically uncombined oxygen in wastewater. Since wastewater or the bacterial degradation of wastes has avidity for oxygen, oxygen dissipated quickly in wastewater. Nitrification contributes to the rapid loss of oxygen within wastewater.

Several important oxygen requirement or tolerant groups of bacteria are recognized (Table 15.1). Significant bacterial groups with respect to their oxygen requirement or tolerance include strict or obligate aerobes, such as nitrifying bacteria and some floc-forming bacteria; facultative anaerobes, such as many organotrophs; and strict or obligate anaerobes, such as methane-forming bacteria.

Strict aerobes use only free molecular oxygen for respiration, while facultative anaerobes use free molecular oxygen for aerobic respiration or substitute another molecule, for example, nitrite ion or nitrate ion, for anaerobic respiration. Many facultative anaerobes also are able to ferment. Fermentation, for example, alcohol production, occurs when bacteria use one organic molecule to degrade another organic molecule.

Oxygen is used in the aeration tank by bacteria for three major

TABLE 15.1 Oxygen Requirement Groups of Bacteria

Group	Respiration	Examples	Significance
Aerobes	Aerobic (O_2)	*Nitrosomonas* *Nitrobacter*	Nitrification
Facultative anaerobes	Aerobic (O_2) or anaerobic	*Bacillus* *Pseudomonas*	cBOD removal Denitrification Floc formation
Strict anaerobes	Anaerobic; killed by O_2	*Methanococcus* *Methanosarcina*	Methane production in anaerobic digesters

biological and operational purposes. These purposes are (1) the oxidation of cBOD to provide carbon and energy for cellular activity, growth, and reproduction (production of MLVSS), (2) oxidation of cBOD to provide energy for endogenous respiration (destruction of MLVSS), and (3) oxidation of nBOD or nitrification.

Of the many operational requirements known to affect nitrifying bacteria or nitrification, dissolved oxygen (DO) concentration is one of the most important requirements. However, an optimal DO concentration to achieve nitrification is relatively low, from 2 to 3 mg/l, and unfortunately, many activated sludge processes are over aerated to achieve nitrification.

The practice of over aeration is not cost-effective and may contribute to shearing of floc particles or enhance foam production. For effective nitrification the amount of DO maintained in the aeration tank should be adjusted to ensure permit compliance and acceptable, mixed liquor effluent concentrations for ammonium ions, nitrite ions, and nitrate ions. In addition to the concentration of DO in the aeration tank, sufficient mixing must be maintained to prevent DO stratification (Figure 15.1), and the DO must penetrate to the core of the floc particles.

Because nitrifying bacteria are strict aerobes, they can only nitrify in the presence of dissolved oxygen (Table 15.2). At DO concentrations <0.5 mg/l, little, if any, nitrification occurs. Factors responsible for this limited amount of nitrification are the lack of oxygen diffusion through the floc particle and competition for oxygen by other organisms. With increasing DO concentration, nitrification accelerates.

Increasing the DO concentration permits better DO penetration

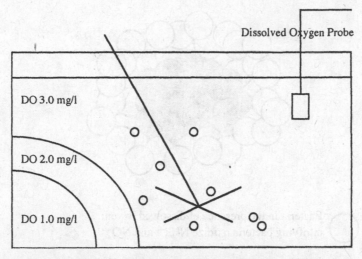

Figure 15.1   *DO stratification. Dissolved oxygen (DO) stratification can occur in an aeration tank due to short-circuiting of flow or lack of cleaning. DO stratification results in zones of varying dissolved oxygen concentration within the aeration tank. DO stratification may cause increased operational costs in order to achieve nitrification or the occurrence of incomplete nitrification. DO stratification can be determine through periodic monitoring of the dissolved oxygen profile of the aeration tank and can be corrected through the use of baffles or routine cleaning of the aeration tank.*

of the floc particle and encourages more nitrification (Figure 15.2). Within the DO range of 0.5 to 1.9 mg/l, nitrification accelerates, but it does not proceed efficiently. Significant nitrification is achieved at DO concentrations from 2.0 to 2.9, while maximum nitrification occurs near a DO concentration of 3.0 mg/l (Figure 15.3). However, if higher DO concentration is maintained in the aeration tank, and cBOD is removed more rapidly due to the higher DO concentration, increased nitrification time will be provided, and additional nitrification can be achieved.

Because nitrifying bacteria must reduce oxidized carbon (CO_2) for cellular growth and reproduction and obtain little energy from the

TABLE 15.2   DO Concentration and Nitrification Achieved

DO Concentration	Nitrification Achieved
<0.5 mg/l	Little, if any, nitrification occurs
0.5 to 1.9 mg/l	Nitrification occurs, but inefficiently
2.0 to 2.9 mg/l	Significant nitrification occurs
3.0 mg/l	Maximum nitrification

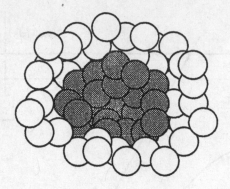

⊘ Bacteria in the presence of dissolved oxyen; nitrifying bacteria oxidize NH_4^+ and NO_2^-

⬤ Bacteria in the absence of dissolved oxyen; nitrifying bacteria do not oxidize NH_4^+ and NO_2^-

Figure 15.2   DO penetration of the floc particle and nitrification. Nitrification occurs only in the presence of free molecular oxygen. As long as dissolved oxygen is present around the perimeter of a floc particle, the bacteria around the perimeter of the floc particle may use the dissolved oxygen to degrade cBOD or oxidize ammonium ions and nitrite ions. As dissolved oxygen penetrates the floc particle, bacteria within the floc particle use the dissolved oxygen and may exhaust any residual dissolved oxygen. In the absence of dissolved oxygen, the bacteria in the core of the floc particle experience an anoxic condition. Also, in the absence of dissolved oxygen, nitrification cannot occur.

oxidation of ammonium ions and nitrite ions, they compete poorly with organotrophs for DO in the aeration tank. Therefore the DO level within the aeration tank should be carefully monitored and not allowed to drop below 1.5 mg/l. Below this value a diminution of nitrifying activity occurs.

Nitrifying bacteria can survive in the absence of DO for only a relatively short period of time. An absence of DO for less than 4 hours does not adversely affect the activity of nitrifying bacteria when DO is restored. An absence of DO for more than 4 hours adversely affects the activity of nitrifying bacteria when DO is restored. An absence of DO for 24 hours or more can destroy the nitrifying bacterial population.

The oxygen demand for complete nitrification is large. For most municipal activated sludge processes the demand will increase the required oxygen supply and power requirement significantly. This

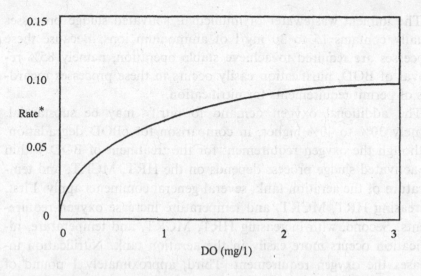

* pound of ammonium ion oxidized per pound MLVSS per day

Figure 15.3 *DO concentration and nitrification. With increasing dissolved oxygen (DO) concentration, the rate of nitrification increases. In laboratory studies, the rate of nitrification eventually levels off at a dissolved oxygen concentration of 3.0 mg/l. However, in an aeration tank, increasing dissolved oxygen concentration above 3.0 mg/l may improve nitrification if the increased dissolved oxygen concentration helps to more rapidly remove cBOD from the aeration tank. With more rapid removal of cBOD in the aeration tank, more time is provided for nitrification. It is the increase in time for nitrification, not the dissolved oxygen concentration, that is responsible for improvement in nitrification.*

increase is due to the consumption of approximately 4.6 pounds of oxygen for each pound of ammonium ions oxidized to nitrate ions (Table 15.3).

The consumption of 4.6 pounds of oxygen for the oxidation of one pound of ammonium ions to one pound of nitrate ions is the theoretical value. The actual or observed amount of oxygen consumed is 4.2 pounds or slightly less than the theoretical value, attributable to the fact that some ammonium ions are not oxidized but are assimilated into cellular material ($C_5H_7NO_2$).

TABLE 15.3   Oxygen Consumption (Theoretical) during Nitrification

Biochemical Reaction	Pounds O_2 Consumed
1 pound NH_4^+ to 1 pound NO_2^-	3.43
1 pound NO_2^- to 1 pound NO_3^-	1.14
1 pound NH^{4+} to 1 pound NO_3^-	4.57

The influent wastewater of municipal, activated sludge processes usually contains 15 to 30 mg/l of ammonium ions. Because these processes are required to achieve stable operation, namely 85% removal of BOD, nitrification easily occurs in these processes regardless of permit requirements for nitrification.

The additional oxygen demand to nitrify may be substantial, namely 30% to 40% higher, in comparison for cBOD degradation. Although the oxygen requirement for the treatment of BOD within an activated sludge process depends on the HRT, MCRT, and temperature of the aeration tank, several general comments apply. First, increasing HRT, MCRT, and temperature increase oxygen requirements. Second, with increasing HRT, MCRT, and temperature, nitrification occurs more easily in the aeration tank. Nitrification increases the oxygen requirement. Third, approximately 1 pound of oxygen is required to oxidize 1 pound of cBOD, and 4.2 pounds of oxygen are required to oxidize 1 pound of nBOD.

16

Alkalinity and pH

Wastewater is normally alkaline. It receives its alkalinity from the potable water supply, infiltration of groundwater, and chemical compounds discharged to the sewer system.

Alkalinity is lost in an activated sludge process during nitrification. This loss occurs through the use of alkalinity as a carbon source by nitrifying bacteria and the destruction of alkalinity by the production of hydrogen ions (H^+) and nitrite ions during nitrification. Hydrogen ions are produced when ammonium ions are oxidized to nitrite ions (Equation 16.1). Significantly more alkalinity is lost through the oxidation of ammonium ions than through the use of alkalinity as a carbon source.

$$NH_4^+ + 1.5O_2 - Nitrosomonas \rightarrow 2H^+ + NO_2^- + 2H_2O \quad (16.1)$$

When hydrogen ions are produced during the oxidation of ammonium ions, nitrous acid (HNO_2) also is produced (Equation 16.2). Nitrous acid destroys alkalinity. The amount of nitrous acid and nitrite ions produced is dependent on the pH of the aeration tank (Figure 16.1).

$$H^+ + NO_2^- \rightarrow HNO_2 \quad (16.2)$$

As alkalinity is lost in the activated sludge process and the pH of the aeration tank drops below 6.7, a significant decrease occurs in

109

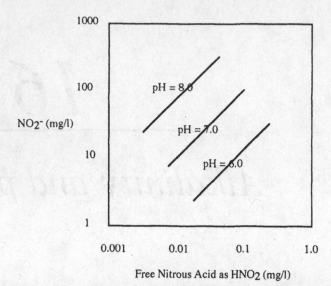

Figure 16.1 *Free nitrous acid and pH. The concentrations of free nitrous acid and the concentration of nitrite ions within an aeration tank are influenced by the pH of the aeration tank. Increasing pH results in a decrease in the concentration of nitrous acid and an increase in the concentration of nitrite ions. Decreasing pH results in an increase in the concentration of nitrous acid and a decrease in the concentration of nitrite ions.*

nitrification (Figure 16.2). Therefore it is important to maintain an adequate amount or residual buffer of alkalinity in the aeration tank to provide pH stability and ensure the presence of inorganic carbon for nitrifying bacteria. The residual amount of alkalinity desired in the aeration tank after complete nitrification is at least 50 mg/l.

Alkalinity refers to those chemicals or alkalis in wastewater that are capable of neutralizing acids. There is a large variety of chemicals in wastewater that provide alkalinity. These chemicals include bicarbonates (HCO_3^-), carbonates (CO_3^{2-}), and hydroxides (OH^-) of calcium, magnesium, and sodium (Table 16.1).

These alkalis provide inorganic carbon (CO_2) for nitrifying bacteria. Nitrifying bacteria prefers bicarbonate alkalinity. When carbon dioxide dissolves in wastewater, it reacts with water to form carbonic acid (H_2CO_3) (Equation 16.3). Carbonic acid dissociates in wastewater to form a hydrogen ion and a bicarbonate ion (Equation 16.4). The bicarbonate ion provides alkalinity.

$$CO_2 + H_2O \rightarrow H_2CO_3 \qquad (16.3)$$

$$H_2CO_3 \leftrightarrow H^+ + HCO_3^- \qquad (16.4)$$

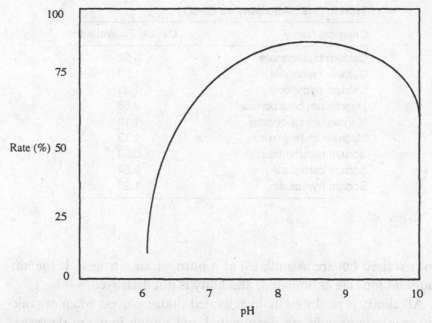

Figure 16.2 *pH and nitrification. With increasing pH, the rate of nitrification increases. The improvement in the rate of nitrification with increasing pH is due to the presence of increased alkalinity and more efficiently operating enzyme systems within the nitrifying bacteria.*

Although alkalinity in wastewater is provided by a variety of chemicals, all chemicals are grouped together, and alkalinity is computed as though the alkalinity is all calcium carbonate (Table 16.2). Approximately 7.14 mg (theoretical) of alkalinity as $CaCO_3$ are destroyed per milligram of ammonium ions oxidized. The actual amount of alkalinity destroyed is 7.07 mg. Some ammonium ions are

TABLE 16.1 Alkalis in Wastewater

Chemical Name	Chemical Formula
Calcium bicarbonate	$Ca(HCO_3)_2$
Calcium carbonate	$CaCO_3$
Calcium hydroxide	$Ca(OH)_2$
Magnesium bicarbonate	$Mg(HCO_3)_2$
Magnesium carbonate	$MgCO_3$
Magnesium hydroxide	$Mg(OH)_2$
Sodium bicarbonate	$NaHCO_3$
Sodium carbonate	Na_2CO_3
Sodium hydroxide	$NaOH$

TABLE 16.2 Alkalinity as CaCO₃

Chemical Name	CaCO₃ Equivalent
Calcium bicarbonate	0.62
Calcium carbonate	1.00
Calcium hydroxide	1.35
Magnesium bicarbonate	0.68
Magnesium carbonate	1.19
Magnesium hydroxide	1.13
Sodium bicarbonate	0.60
Sodium carbonate	0.94
Sodium hydroxide	1.25

not oxidized but are assimilated as a nutrient for nitrogen. If the ammonium ions are assimilated, alkalinity is not destroyed.

Alkalinity is produced in an activated sludge process when organic-nitrogen compounds are deaminated and nitrate ions are destroyed during denitrification. The amount of alkalinity produced or returned to an activated sludge process during denitrification is 3.57 mg as CaCO₃ per milligram of nitrate ions that are reduced to molecular nitrogen. This amount of alkalinity that is returned during denitrification is approximately one-half the amount of alkalinity that is lost during nitrification. Therefore the net alkalinity change in an activated sludge process through bacterial activity is a function of

- organic-nitrogen compounds deaminated,
- ammonium ions converted to nitrite ions,
- ammonium ions assimilated into new cells or MLVSS, and
- nitrate ions destroyed during denitrification.

It is essential for successful nitrification that an activated sludge process be adequately buffered with alkalinity to counteract its tendency to become more acidic over time through nitrification. To ensure that an adequate amount of alkalinity is maintained in the aeration tank during nitrification, a residual concentration or target value of at least 50 mg/l of alkalinity is recommended after complete nitrification. If this value for alkalinity is not present, then alkalinity should be added to the aeration tank.

Although numerous chemicals may be used to add alkalinity, commonly used chemicals for alkalinity addition are listed in Table 16.3.

TABLE 16.3 Alkalis Suitable for Alkalinity Addition

Chemical Name	Formula	Common Name
Sodium bicarbonate	$NaHCO_3$	Baking soda
Calcium carbonate	$CaCO_3$	Calcite
		Limestone
		Whiting chalk
Sodium carbonate	Na_2CO_3	Soda ash
Calcium hydroxide	$Ca(OH)_2$	Lime
Sodium hydroxide, 50%	$NaOH$	Caustic soda

The addition of alkalinity to the aeration tank through the use of calcium carbonate is shown in Equation 16.5.

$$H_2O + CO_2 + CaCO_3 \rightarrow Ca(HCO_3)_2 \leftrightarrow Ca^{2+} + 2HCO_3^- \quad (16.5)$$

Aside from temperature, the hydrogen ion concentration or pH of an organism's environment exerts the greatest influence upon the organism. Nitrification proceeds much more slowly at low pH, and it is likely that in most environments, nitrification below pH 5.0 is not due to nitrifying bacteria but to organotrophs, including fungi. At neutral pH values nitrifying bacteria are dominant, and at alkaline pH values nitrification is due mostly, if not entirely, to nitrifying bacteria.

Low pH in wastewater has a primary effect on nitrifying bacteria by inhibiting enzymatic activity and a secondary effect on the availability of alkalinity. Nitrification in an activated sludge process begins to accelerate above pH 6.7 (Table 16.4), and the optimal pH range for nitrification is 7.2 to 8.0. At the pH range of 7.2 to 8.0 the rate of nitrification is assumed to be constant, and many activated

TABLE 16.4 pH and Nitrification

pH	Impact upon Nitrification
4.0 to 4.9	Nitrifying bacteria present; organotrophic nitrification occurs
5.0 to 6.7	Nitrification by nitrifying bacteria; rate of nitrification sluggish
6.7 to 7.2	Nitrification by nitrifying bacteria; rate of nitrification increases
7.2 to 8.0	Nitrification by nitrifying bacteria; rate of nitrification assumed constant
7.5 to 8.5	Nitrification by nitrifying bacteria

sludge processes nitrify at a pH close to neutral. Although a higher pH would appear to be more desirable for nitrification, the higher pH would adversely affect many organotrophs that are required to degrade cBOD. Fortunately nitrifying bacteria are able to slowly acclimate to a pH less than optimal. However, this acclimation may require a gradual increase or decrease of pH. The pH at which nitrifying bacteria acclimate must be maintained at a steady-state condition.

17

Temperature

Of all operational factors affecting nitrification, temperature has the most significant influence on the growth of nitrifying bacteria and, consequently, the rate of nitrification (Table 17.1). The rate of nitrification usually is expressed as pounds of ammonium ions oxidized per pound of MLVSS per day. Because nitrifying bacteria are temperature sensitive, nitrification is temperature sensitive. There is a significant reduction in the rate of nitrification with decreasing temperature and, conversely, a significant acceleration in the rate of nitrification with increasing temperature. The rate of growth of nitrifying bacteria increases considerably with temperature over the range of 8° to 30°C, with *Nitrosomonas* having nearly a 10% increase in growth rate per 1°C rise.

Below 10°C the nitrification rate sharply falls. Above 10°C the rate of nitrification is almost directly proportional to the temperature. *Nitrosomonas* isolated from activated sludge processes has an

TABLE 17.1 Temperature and Nitrification

Temperature	Effect upon Nitrification
>45°C	Nitrification ceases
28° to 32°C	Optimal temperature range
16°C	Approximately 50% of nitrification rate at 30°C
10°C	Significant reduction in rate, approximately 20% of rate at 30°C
<5°C	Nitrification ceases

TABLE 17.2 Temperature and MCRT Required for Nitrification

Temperature	MCRT
10 °C	30 days
15 °C	20 days
20 °C	15 days
25 °C	10 days
30 °C	7 days

optimal growth rate at 30 °C. Therefore, for operational purposes, the optimal temperature for nitrification in the activated sludge process is generally considered to be 30 °C. No growth of *Nitrosomonas* or *Nitrobacter* occurs below 4 °C.

Because of the decreased activity and reproduction of nitrifying bacteria during cold temperatures, an increase in the size of nitrifying bacteria (MLVSS) or an increase in MCRT is required in order to maintain effective nitrification (Table 17.2 and Figure 17.1). In gen-

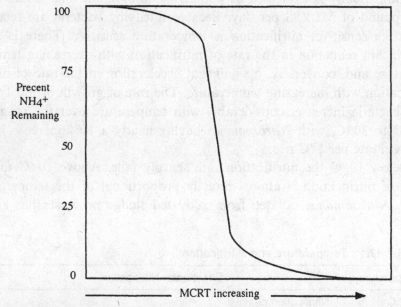

Figure 17.1 *MCRT and nitrification. With increasing mean cell residence time (MCRT), more solids including nitrifying bacteria are retained in the aeration tank and more time is provided for nitrifying bacteria to reproduce. Therefore increasing MCRT results in rapid removal or oxidation of ammonium ions in the aeration tank.*

eral, the shift to nitrification is a shift to an older sludge or biomass. A MCRT of 10 days or more is usually required to nitrify effectively. A low temperature may increase the time required to nitrify.

Since nitrifying bacteria are strict aerobes, they are found only in the top one to two inches of the soil. When the ground freezes below two inches, a significant source of "seeding" of nitrifying bacteria to the activated sludge process is lost. Due to decreased activity and growth in nitrifying bacteria and loss of a significant "seed" source during cold temperatures, many regulatory agencies in temperate regions provide seasonally adjusted nitrification requirements, that is, seasonally adjusted ammonia discharge limits. For example, a higher discharge limit for ammonia is provided during cold temperature months, October through April, and a lower discharge limit for ammonia is provided during warm temperature months, May through September.

The inhibitory effect of cold temperature is greater on *Nitrobacter* than *Nitrosomonas*. Therefore it is not uncommon for nitrite ions to accumulate during cold temperatures.

18

Inhibition and Toxicity

There are several forms of inhibition and toxicity that may occur during nitrification (Table 18.1). Inhibition is a temporary, short-term (acute) or long-term (chronic) loss of enzymatic activity. Toxicity is the permanent loss of enzymatic activity or irreversible damage to cellular structure.

Although nitrifying bacteria can overcome (acclimate) to inhibition by repairing damaged enzyme systems, chronic inhibition may significantly lower the rate of reproduction of nitrifying bacteria, resulting in a "washout" of the population through the loss of bacteria in the secondary effluent or sludge wasting. Also nitrifying bacteria obtain only a relatively small amount of energy from the oxidation of ammonium ions and nitrite ions, and unfortunately, sufficient energy is often not available for sustained or repeated acclimation. Therefore relatively small increases of inhibitory wastes can cause dramatic changes in the growth of nitrifying bacteria.

Nitrifying bacteria cannot acclimate as often to an inhibitory event as can organotrophs. This is due to the fact that organotrophs obtain a relatively large amount of energy from the oxidation of cBOD, and it is energy that permits them to repair damaged enzyme systems. Therefore nitrifying bacteria are more "sensitive" to inhibitory wastes than organotrophs. Thus significant loss of nitrification occurs before significant loss in efficiency of cBOD removal.

Due to the relatively small amount of energy available for acclimation, nitrifying bacteria are "sensitive" to very low concentrations

TABLE 18.1   Forms of Inhibition or Toxicity

Form of Inhibition or Toxicity	Description or Example
Free chlorine residual	Chlorination of RAS and effluent
Inorganic	Heavy metals and cyanide
Organic	Industrial wastes, e.g., phenol
pH	<5.0
Substrate	Free ammonia and free nitrous acid
Sunlight	Ultraviolet radiation
Temperature	<5°C

of inorganic wastes (Table 18.2) and organic wastes (Table 18.3). Wastes that are highly toxic to nitrifying bacteria include cyanide, halogenated compounds, heavy metals, hydantoins, mercaptans, phenols, and thiourea.

Nitrifying bacteria also are inhibited by relatively low concentrations of free ammonia and free nitrous acid. Free ammonia is produced from ammonium ions under a high pH in the aeration tank. Free nitrous acid is produced from nitrite ions under a low pH in the aeration tank. This inhibition or toxicity to free ammonia and free nitrous acid is known as substrate inhibition or toxicity.

The conversion of ammonium ions to free ammonia and its accumulation are a function of ammonium ion concentration and aeration tank pH. With increasing pH, ammonium ions are more easily converted to free ammonia (Equation 18.1).

$$NH_4^+ + OH^- \leftrightarrow NH_3 + H_2O \qquad (18.1)$$

TABLE 18.2   Inhibitory Threshold Concentrations of Some Inorganic Wastes

Inorganic Waste	Concentration, mg/l
Chromium (hexavalent)	0.25
Chromium (trivalent)	0.05
Copper	0.35
Cyanide	0.50
Mercury	0.25
Nickel	0.25
Silver	0.25
Sulfate	500
Zinc	0.30

TABLE 18.3   Inhibitory Threshold Concentrations of Some Organic Wastes

Organic Waste	Concentration, mg/l
Allyl alcohol	20.0
Aniline	8.0
Chloroform	18.0
Mercaptobenzothiazole	3.0
Phenol	6.0
Skatol	7.0
Thioacetamide	0.5
Thiourea	0.1

Free ammonia inhibits *Nitrosomonas* and *Nitrobacter*. Free ammonia can inhibit *Nitrosomonas* at concentrations as low as 10 mg/l. Free ammonia can inhibit *Nitrobacter* at concentrations as low as 0.1 mg/l.

The conversion of nitrite ions to free nitrous acid and its accumulation are a function of nitrite ion concentration and aeration tank pH. With decreasing pH, nitrite ions are more easily converted to free nitrous acid (Equation 18.2).

$$NO_2^- + H^+ \leftrightarrow NHO_2 \qquad (18.2)$$

Free nitrous acid inhibits *Nitrosomonas* and *Nitrobacter* at very low concentrations. Both genera of bacteria may be inhibited by free nitrous acid at concentrations as low as 1.0 mg/l.

Substrate inhibition in an activated sludge process usually occurs at a concentration of 400 to 500 mg/l ammonium ions or when ammonium ions are converted to nitrite ions at a faster rate than nitrite ions are converted to nitrate ions. Therefore excessive ammonium ion discharge or deamination of organic-nitrogen compounds may inhibit nitrification.

Inhibition due to high ammonium ion concentration or high organic-nitrogen levels can be prevented. Reducing or equalizing nitrogenous waste discharges to the activated sludge process helps to prevent substrate inhibition or toxicity, and maintaining proper pH and alkalinity in the aeration tank also helps to prevent substrate inhibition or toxicity.

Except for the photosynthetic bacteria, ultraviolet radiation or light harms most bacteria, including nitrifying bacteria. The lethal

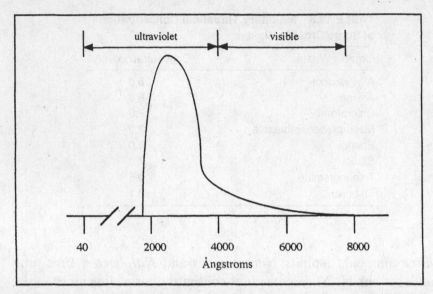

Figure 18.1 *Ultraviolet radiation. Ultraviolet radiation from approximately 2000 to 4000 Ångstroms in length is highly lethal to nitrifying bacteria. The lethal effect of ultraviolet radiation upon nitrifying bacteria prevents their growth on the surface of soil and the surface of biofilm that coats aeration tanks and secondary clarifiers.*

effects of ultraviolet radiation are limited to only short, invisible wavelengths of light (Figure 18.1). The most harmful wavelength is 2650 Ångstroms (265 nm or 0.265 μm). It is suspected that ultraviolet radiation causes inactivation of enzyme systems, especially in young or rapidly growing cells.

19

Mode of Operation

The mode of operation of an activated sludge process addresses those critical factors that have significant influence on the population size of nitrifying bacteria and the ability of the process to nitrify. Those operational factors that influence nitrification most in an activated sludge process are MCRT, MLVSS, HRT, F/M, and ammonium ion concentration (Table 19.1)

MCRT AND MLVSS

In order to establish a large population of nitrifying bacteria, most activated sludge processes operate at a relatively high MCRT. The minimum MCRT required to nitrify is affected by several operational factors, especially temperature. However, the MCRT needed to achieve significant nitrification is usually two to three times the generation time of nitrifying bacteria. The generation time of nitrifying bacteria in an activated sludge process is considered to be two to three days.

Because temperature affects biological activity and generation time of nitrifying bacteria, the MCRT, MLVSS, and HRT required to achieve complete nitrification is inversely related to temperature (Table 19.2). With increasing temperature, increased biological activity and shorter generation times occur for nitrifying bacteria. Therefore a lower MCRT, lower MLVSS and smaller HRT are required. The reverse is true for colder temperature.

123

TABLE 19.1 Operational Factors Favoring Nitrification

Operational Factor	Range/Value
MCRT	>8 days (increasing with decreasing temperature)
MLVSS	>2000 mg/l
HRT	>10 hours during cold temperatures
F/M	0.5 pounds ammonium ions per pound MLVSS
Ammonium ion	<400 mg/l

There are several advantages of maintaining as many nitrifying bacteria as possible. First, having more nitrifying bacteria in the aeration tank can offset the reduction in nitrification due to cold temperature. Second, by increasing the MCRT to provide for more nitrifying bacteria, the increased MCRT also provides for more organotrophs and more rapid removal of soluble cBOD. The rapid removal of soluble cBOD provides for more aeration time for nitrification. Third, a decrease in "sensitivity" to inhibiting and toxic wastes is provided for nitrification by maintaining more nitrifying bacteria and organotrophs.

By increasing the MCRT and MLVSS, the inhibitory or toxic mass to biomass ration is lowered; that is, the ratio of the quantity of inhibitory or toxic wastes to the quantity of nitrifying bacteria is lowered. The lower ratio provides for more viable nitrifying bacteria after a discharge of inhibitory or toxic wastes to the aeration tank.

There are several disadvantages of maintaining as many nitrifying bacteria as possible. First, increasing the quantity of solids discharged to the secondary clarifiers may contribute to poor settling of solids in the clarifier and loss of solids from the clarifier. Second, rapid depletion of DO may occur due to the presence of rapid DO consumption by a relatively large population of bacteria. Third, denitrification may occur in the secondary clarifier.

TABLE 19.2 Temperature, MLVSS, and Aeration Time Required to Completely Nitrify

MLVSS	Temperature	
	Aeration Time at 12°C	Aeration Time at 17°C
2000 mg/l	8 to 16 hours	6 to 12 hours
4000 mg/l	6 to 9 hours	4 to 5 hours

Note: Moderate strength BOD used (200 mg/l).

HRT

In addition to the presence of an adequate number of nitrifying bacteria and organotrophs, an adequate HRT in the aeration tank must be provided. A high HRT is required for organotrophs to efficiently remove cBOD and for nitrifying bacteria to oxidize ammonium ions and nitrite ions. Increased HRT may be achieved in the activated sludge process by (1) placing more aeration tanks in service, (2) thickening solids in the secondary clarifier and reducing the RAS rate, and (3) reducing inflow and infiltration.

Increasing HRT provides more efficient cBOD removal and more time for nitrification. During cold temperatures, an HRT of at least 10 hours may be required to nitrify.

As HRT is increased, more particulate BOD is solublized. The increased presence of soluble cBOD places an increased oxygen demand upon the aeration tank. The increased oxygen demand may hinder nitrification. Therefore improved particulate BOD removal in the primary clarifiers may be required to maintain efficient nitrification at an increased HRT value.

There are three means of improving primary clarifier efficiency. First, more primary clarifiers may be placed in service. Second, settled solids may be removed more quickly, and third, a polymer or metal salt (coagulant), such as lime, may be added to the primary clarifier influent to remove additional solids.

F/M

The maintenance of a low F/M of sufficient duration provides a relatively long time for the nitrifying bacteria to increase significantly in number. The effect of F/M on nBOD removal efficiency is shown in Figure 19.1. In terms of ammonium ions (food or "F") to MLVSS (microorganisms or "M") ratio, a threshold of 0.5 pounds of ammonium ions applied per day per pound of MLVSS may be applicable.

AMMONIUM ION CONCENTRATION

Excess ammonium ion concentration as well as excess cBOD adversely affects the growth of nitrifying bacteria. At high pH and low pH, excess ammonium ions may contribute to substrate inhibition or toxicity. Excess cBOD causes a significant oxygen demand. This de-

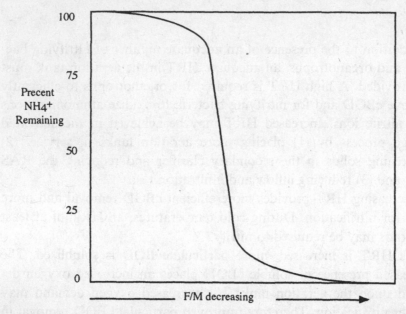

Figure 19.1 *F/M and nitrification. With decreasing F/M the amount of substrate or food is decreased while the number of bacteria is increased. With decreasing amounts of substrate and increasing numbers of bacteria, the substrate (cBOD or nBOD) is more rapidly removed or oxidized. Therefore decreasing F/M results in more rapid and efficient removal or oxidation of ammonium ions and nitrite ions.*

mand may cause a drop in DO that adversely affects nitrifying bacteria.

The rapid loss of DO within an aeration tank "starves" the nitrifying bacteria of essential DO. Due to fluctuations in cBOD loading, nitrification may occur intermittently in an activated sludge process.

The impact of cBOD loading upon nitrification can be observed by monitoring aeration tank influent cBOD/TKN. The rate of nitrification increases as the cBOD/TKN decreases. The impact of cBOD loading also can be observed by monitoring the aeration tank influent cBOD and MLVSS. Most municipal activated sludge processes experience a decrease in the rate of nitrification when the cBOD exceeds 0.5 pounds per pound of MLVSS.

The adverse impacts of excess ammonium ion concentration and excess cBOD loading may be prevented by (1) equalizing flow from dischargers of industrial wastes and (2) requiring dischargers of industrial wastes to provide the earliest possible notification of significant loading changes. By providing the earliest possible notice, more time would be made available to implement necessary process control measures.

20

Classification of
Nitrification Systems

Nitrification systems within activated sludge processes may be classified according to the degree of separation of cBOD and nBOD oxidation or removal. The removal of cBOD and nBOD may occur in the same aeration tank or in separate aeration tanks. If cBOD and nBOD removal occur in the same aeration tank, the treatment system is termed a "single-stage" system (Figure 20.1). If cBOD and nBOD removal occur in separate aeration tanks, the treatment system is termed a "separate-stage" or "two-stage" system (Figure 20.2). The activated sludge processes most commonly used for nitrification are complete mix, plug-flow, extended aeration, and the oxidation ditch.

Although nBOD removal can be accomplished in the same aeration tank as is used for cBOD removal (single-stage system), nitrification in a separate tank (two-stage system) allows greater process

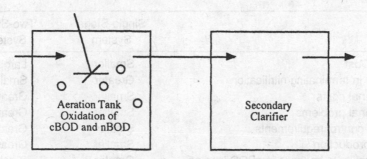

Figure 20.1 *Single-stage nitrification system. In a single-stage nitrification system, an aeration tank or a series of aeration tanks is used to oxidize cBOD and nBOD. In a single-stage system, oxidation of cBOD and oxidation of nBOD occur in the same aeration tank.*

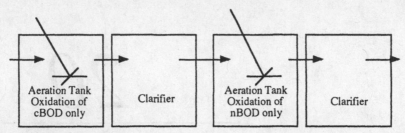

Figure 20.2 *Two-stage nitrification system. In a two-stage nitrification system, an aeration tank or series of aeration tanks is used to oxidize cBOD; another aeration tank or series of aeration tanks is used to oxidize nBOD. In a two-stage system, oxidation of cBOD and oxidation of nBOD occur in different aeration tanks.*

control. In a two-stage system each aeration tank can be operated independently, and many wastes that are inhibitory or toxic to nitrifying bacteria may be removed or reduced in concentration in the first aeration tank or cBOD removal stage. It is commonly accepted that a cBOD of 40 to 50 mg/l fed to the second aeration tank or nitrification tank can be tolerated by nitrifying bacteria.

The system used for nitrification depends on a number of factors including (1) plant design, (2) whether seasonal or year-round ammonia discharges requirements are to be satisfied, (3) range of temperatures, (4) desired effluent concentration for ammonium ions, (5) other effluent, water quality requirements, and (6) costs. A comparison of advantages and disadvantages of single-stage and two-stage nitrification systems is presented in Table 20.1. Due to the difficulty in achieving complete nitrification year round in more northern geographic locations or temperate regions, it may become mandatory to accomplish nitrification in a two-stage system.

TABLE 20.1 Comparison of Single-Stage and Two-Stage Nitrification Systems

Factor	Single-Stage System	Two-Stage System
Capital Cost	Smaller	Larger
Difficulty in terminating nitrification	Greater	Smaller
Operational costs	Smaller	Greater
Operational problems	Smaller	Greater
Process control requirements	Smaller	Greater
Sludge production	Smaller	Greater
Susceptibility to changes in nBOD loading	Greater	Smaller
Susceptibility to changes in toxic loading	Greater	Smaller

21

Troubleshooting Key and Tables

To identify the factor responsible for the loss of nitrification in an activated sludge system, answer question 1 below in the troubleshooting key. Continuing answering each question, until "yes" is obtained. Proceed with a review of the table or corrective measure indicated by the "yes."

1. Is the wastewater temperature < 16°C?
   Yes: See Table 21.1.
   No: Go to question 2.

TABLE 21.1 Corrective Measures for Cold Wastewater Temperature

Corrective Measure	√ If Applicable
Ensure adequate alkalinity	
Increase dissolved oxygen concentration	
Increase HRT	
Increase MCRT/MLVSS	
Use bioaugmentation products	
Use biological holdfast system	

2. Is the dissolved oxygen in the aeration tank < 1.5 mg/l?
   Yes: See Table 21.2.
   No: Go to question 3.

TABLE 21.2   Corrective Measures for Low Dissolved Oxygen

Corrective Measure	√ If Applicable
Equalize flow	
Identify and reduce/remove O_2 scavengers	
Reduce pBOD loading	
Troubleshoot and repair aeration equipment	

3. Is the alkalinity in the ML effluent < 50 mg/l after complete nitrification?

   Yes: Adequate alkalinity is not available, add alkalinity.

   No: Go to question 4.

4. Is MCRT or MLVSS less than normal?

   Yes: Increase MCRT or MLVSS.

   No: Go to question 5.

5. Are inhibitory wastes present?

   Yes: See Table 21.3.

   No: Go to question 6.

TABLE 21.3   Corrective Measures for Inhibitory Wastes

Corrective Measure	√ If Applicable
Identify and regulate industrial waste	
Increase MCRT and MLVSS	
Use bioaugmentation products	
Use biological holdfast system	

6. Is the ammonium ion concentration or organic-nitrogen loading high?

   Yes: See Table 21.4.

   No: Go to question 7.

TABLE 21.4 Corrective Measures for High-Substrate (Ammonium Ion) Loading

Corrective Measure	✓ If Applicable
Ensure acceptable mixed liquor pH	
Equalize ammonium ion loading	
Equalize organic-nitrogen loading	

7. Is the cBOD concentration in the aeration tank effluent filtrate high?

   Yes: See Table 21.5.

   No: Go to question 8.

TABLE 21.5 Corrective Measures for Elevated cBOD Loading

Corrective Measure	✓ If Applicable
Decrease pBOD loading	
Decrease fat, oil, and grease loading	
Increase dissolved oxygen concentration	
Increase HRT	
Increase MCRT and MLVSS	
Use bioaugmentation products	
Use biological holdfast system	

8. Is there a significant decrease in HRT?

   Yes: See Table 21.6.

   No: Go to question 9.

TABLE 21.6 Corrective Measures for Inadequate HRT

Corrective Measure	✓ If Applicable
Correct significant sources of I/I	
Place additional aeration tanks in service	
Reduce RAS rate	
Thicken clarifier solids and reduce RAS rate	

9. Is there a significant increase in HRT?

    Yes: See Table 21.7.

    No: Go to question 10.

TABLE 21.7   Corrective Measures for Excessive HRT

Corrective Measure	√ If Applicable
Increase pBOD loading	
Remove aeration tanks from service	

10. Is there a decrease in efficiency of pBOD removal in the primary clarifier?

    Yes: See Table 21.8

    No: See question 11.

TABLE 21.8   Corrective Measures for Increased pBOD Loading

Corrective Measure	√ If Applicable
Add metal salt or polymer	
Place additional clarifier in service	
Remove clarifier solids more frequently	

11. Perform ammonium ion, nitrite ion, and nitrate ion testing of the mixed liquor effluent to determine if the loss of nitrification is due to depressed temperature or a limiting process condition.

Part III

Denitrification

22

Introduction to Denitrification

The term "denitrification" was first used in France in 1886 to describe the use of nitrate ions by some bacteria to degrade substrate. The bacterial use of nitrate ions (and nitrite ions) to degrade substrate actually evolved before the use of free molecular oxygen.

Wastewater denitrification describes the use of nitrite ions or nitrate ions by facultative anaerobes (denitrifying bacteria) to degrade cBOD. Although denitrification often is combined with aerobic nitrification to remove various forms of nitrogenous compounds from wastewater, denitrification occurs whenever an anoxic condition exists. Therefore denitrification can promote favorable operational conditions or can contribute to operational problems.

Facultative anaerobes make up approximately 80% of the bacteria within an activated sludge process. These organisms have the enzymatic ability to use free molecular oxygen, nitrite ions, or nitrate ions to degrade cBOD. Facultative anaerobes prefer and use free molecular oxygen when it is available. The use of free molecular oxygen provides the bacteria with more energy for cellular activity, growth, and reproduction than does the use of nitrite ions or nitrate ions.

Bacterial degradation of cBOD is "respiration." Respiration may be aerobic (oxic) or anaerobic. Aerobic respiration occurs when free molecular oxygen is available and is used to degrade cBOD, such as glucose ($C_6H_{12}O_6$) (Equation 22.1).

$$C_6H_{12}O_6 + O_2 \rightarrow 6CO_2 + 6H_2O \qquad (22.1)$$

135

Anaerobic respiration occurs when free molecular oxygen is not available and another molecule is used to degrade cBOD. Molecules other than free molecular oxygen that can be used to degrade cBOD include nitrite ions and nitrate ions. The molecule used for the degradation of cBOD is dependent on its availability, the presence of other molecules, and the enzymatic ability of the bacterial population. If nitrite ions or nitrate ions are used to degrade cBOD, such as a five-carbon sugar, this form of respiration is termed "anoxic" (Equation 22.2).

$$C_5H_{10}O_5 + 4NO_3^- + 4H^+ \rightarrow 5CO_2 + 7H_2O + 2N_2 \quad (22.2)$$

During anoxic respiration, nitrite ions and nitrate ions are reduced (oxygen removed from the ions) through several biochemical steps or reactions. The principle gaseous end product of the biochemical reactions is molecular nitrogen.

Anoxic respiration or denitrification is termed "dissimilatory" nitrite or nitrate reduction, because nitrite ions and nitrate ions, respectively, are reduced to from molecular nitrogen. The nitrogen in the nitrite ions or nitrate ions is not incorporated into cellular material, the nitrogen in the ions is loss to the atmosphere as a gas.

Nitrification does not remove nitrogen from the wastewater, it simply transforms it from ammonium ions to nitrate ions. Denitrification removes nitrogen from the wastewater by converting it to insoluble gases that escapes to the atmosphere. Besides molecular nitrogen, nitrous oxide (N_2O) is produced during denitrification from nitrite ions and nitrate ions. This nitrogen-containing gas is insoluble in wastewater and escapes to the atmosphere.

When nitrite ions and nitrate ions are reduced to ammonium ions inside the bacterial cell, the nitrogen in the ammonium ions is incorporated into cellular material. This reduction of nitrogen is termed "assimilatory" nitrite or nitrate reduction.

Assimilatory nitrite reduction and assimilatory nitrate reduction do not remove nitrogen from the wastewater.

23

Denitrifying Bacteria

Although several groups of organisms are capable of denitrification, including fungi and the protozoa *Loxodes*, most denitrifying organisms consist of facultative anaerobic bacteria. The bacteria that denitrify are known by several names including denitrifiers, heterotrophs, and organotrophs.

Denitrifying bacteria degrade cBOD using nitrite ions and nitrate ions in the absence of free molecular oxygen. The bacteria degrade cBOD in order to obtain energy for cellular activity and carbon for cellular synthesis (growth and reproduction).

A relatively large number of genera of facultative anaerobes are capable of denitrification (Table 23.1). Most denitrifiers reduce nitrate ions via nitrite ions to molecular nitrogen without the accumulation of intermediates. However, some denitrifiers lack key enzyme systems to denitrify completely, and the lack of these enzyme systems does permit the production and accumulation of intermediates.

Although there are numerous genera of denitrifying bacteria, all denitrifying genera do not contain large numbers of species, and all denitrifying bacteria do not respire similarly. The genera *Alcaligenes*, *Bacillus*, and *Pseudomonas* contain the largest number of denitrifying bacteria.

Many genera of denitrifying bacteria can use nitrite ions or nitrate ions to degrade cBOD, some genera such as *Enterobacter* and *Escherichia* can use only nitrate ions. Other genera such as *Alcaligenes* can use only nitrite ions. The use of nitrate ions in this manner is known as nitrate respiration, while the use of nitrite ions is known as

TABLE 23.1   Genera of Bacteria That Include Denitrifying Species

Acetobacter	Halobacterium
Achromobacter	Hyphomicrobium
Acinetobacter	Kingella
Agrobacterium	Methanonas
Alcaligenes	Moraxella
Arthrobacter	Neisseria
Axotobacter	Paracoccus
Bacillus	Propionicbacterium
Chromobacterium	Pseudomonas
Corynebacterium	Rhizobium
Denitrobacillus	Rhodopseudomonas
Enterobacter	Spirillum
Escherichia	Thiobacillus
Flavobacterium	Xanthomonas

nitrite respiration. The reduction of nitrate ions to only nitrite ions during denitrification may result in an accumulation of nitrite ions. This form of respiration in a secondary clarifier may result in the production of the "chlorine sponge."

Some genera of denitrifying bacteria are microaerophillic and can tolerate only low levels of free molecular oxygen. Some genera of denitrifying bacteria including species of *Corynebacterium* and *Pseudomonas* do not denitrify completely and produce nitrous oxide instead of molecular nitrogen as their gaseous end product.

Most denitrifying bacteria cannot ferment, that is, use a molecule of cBOD to degrade another molecule of cBOD. However, some species of *Bacillus* and *Chromobacterium* can denitrify and ferment at the same time. Finally, species of *Propionicbacterium* that denitrify cannot respire aerobically, that is, cannot use free molecular oxygen.

The enzymatic machinery needed for denitrification is formed only under an anoxic condition or the presence of a low oxygen concentration. However, the production of the enzymatic machinery for denitrification is accomplished quickly.

Denitrifying bacteria are common soil and water organisms and are associated with fecal waste. Denitrifying bacteria enter an activated sludge process through I/I and domestic wastewater. Many denitrifying bacteria are floc-forming organisms or are easily adsorbed to floc particles. Most denitrifying bacteria reproduce every 15 to 30 minutes and are present in an activated sludge process in millions per milliliter of bulk solution and billions per gram of MLVSS.

24

Biochemical Pathway and Respiration

The biochemical pathway for denitrification refers to the sequential steps of chemical reactions occurring inside the bacterial cells as nitrite ions and nitrate ions are reduced to molecular nitrogen during cBOD degradation. The overall degradation of cBOD using nitrate ions can be expressed in two, simplistic biochemical reactions (Equations 24.1 and 24.2).

$$NO_3^- + cBOD \rightarrow NO_2^- + CO_2 + H_2O \qquad (24.1)$$

$$NO_2^- + cBOD \rightarrow N_2 + CO_2 + H_2O \qquad (24.2)$$

The biochemical pathway for denitrification involves a stepwise conversion of nitrate ions to molecular nitrogen. These steps are the conversion of nitrate ions to nitrite ions, the conversion of nitrite ions to nitric oxide (NO), the conversion of nitric oxide to nitrous oxide (N_2O), and the conversion of nitrous oxide to molecular nitrogen.

There are five significant, nitrogenous compounds involved in denitrification (Table 24.1). The nitrate ion is considered to be the initial substrate for denitrification, and molecular nitrogen is considered to be the product from denitrification. Nitrite ions, nitric oxide, and nitrous oxide are considered to be the intermediates. The formation of the intermediates and their release to the wastewater is enhanced during changes between aerobic and anoxic conditions.

Although nitric oxide is considered to be an intermediate, it may be in equilibrium with the two intermediates, nitrite ions and nitrous

TABLE 24.1   Nitrogenous Compounds Involved in Denitrification

Nitrogenous Compound	Formula
Nitrate ion	NO_3^-
Nitrite ion	NO_2^-
Nitric oxide	NO
Nitrous oxide	N_2O
Molecular nitrogen	N_2

oxide. Some denitrifying bacteria release all three intermediates during denitrification, while other bacteria release two, one, or none of the intermediates.

In the presence of adequate organic carbon or cBOD and absence of free molecular oxygen, biological denitrification can occur. Adequate organic carbon is considered to be a soluble cBOD to nitrite ion and nitrate ion ratio of approximately 3:1.

There are two bacterial, energy yielding steps or reactions involved in denitrification that not only provide the bacteria with energy but also release molecular nitrogen to the atmosphere. These two energy-yielding reactions and the overall energy-yielding reaction are provided in Equations (24.3), (24.4), and (24.5). These equations use methanol as the cBOD source.

$$6NO_3^- + 2CH_3OH \rightarrow 6NO_2^- + 2CO_2 + 4H_2O \qquad (24.3)$$

$$6NO_2^- + 3CH_3OH \rightarrow 3N_2 + 3CO_2 + 3H_2O + 6OH^- \qquad (24.4)$$

$$6NO_3^- + 5CH_3OH \rightarrow 3N_2 + 5CO_2 + 7H_2O + 6OH^- \qquad (24.5)$$
$$\text{(overall energy-yielding reaction)}$$

Because energy obtained through aerobic respiration of cBOD is greater than the energy obtained through anoxic respiration of cBOD, denitrifying bacteria prefer aerobic respiration or the use of free molecule oxygen to degrade cBOD. Therefore, in the presence of a high dissolved oxygen concentration (> 1.0 mg/l), denitrifying bacteria activate their enzymatic machinery for using free molecular oxygen and deactivate their enzymatic machinery for using nitrite ions and nitrate ions. However, the energy obtained from anoxic respiration compares well with aerobic respiration (Equations 24.6 and 24.7).

$$\text{Glucose} + 6O_2 \rightarrow 6CO_2 + 6H_2O + 686 \text{ kcal} \qquad (24.6)$$
$$\text{(aerobic respiration)}$$

$$\text{Glucose} + 4.8NO_3^- + 4.8H^+$$
$$\rightarrow 6CO_2 + 2.4N_2 + 8.4H_2O \qquad + 636 \text{ kcal} \qquad (24.7)$$
$$\text{(anoxic respiration)}$$

Like aerobic respiration, denitrification allows a complete oxidation of the organic substrate (cBOD) to carbon dioxide. In aerobic respiration, free molecular oxygen serves as the final electron carrier molecule. In anoxic respiration, nitrite ions or nitrate ions serve as the final electron carrier molecule.

Approximately 25% of the cBOD degraded under anoxic respiration is used for cellular synthesis or sludge production (Equation 24.8). The quantity of cells or sludge produced under aerobic respiration is greater due to the large quantity of energy obtained through aerobic respiration as compared to anoxic respiration.

$$NO_3^- + 1.8CH_3OH + H^+$$
$$\rightarrow 0.065C_5H_7O_2N^* + 0.47N_2 + 0.76CO_2 + 2.44H_2O \quad (24.8)$$
$$(^* \text{ new cells or sludge})$$

The hydroxyl ion (OH^-) and some of the carbon dioxide produced during denitrification are returned to the activated sludge process as alkalinity. This return is important because much alkalinity is lost in the activated sludge process during nitrification. Approximately 50% of the alkalinity lost during nitrification is returned during denitrification.

25

Gaseous End Products

The compounds resulting from a biochemical reaction are known as products (Equation 25.1). When denitrification occurs, the denitrifying bacteria form several gaseous products. These products formed during denitrification include molecular nitrogen, carbon dioxide, nitrous oxide, ammonia, and nitric oxide.

$$\text{Reactants} \rightarrow \text{Products} \qquad (25.1)$$

The majority of the gases formed and released by denitrifying bacteria consists of molecular nitrogen and carbon dioxide. Molecular nitrogen is insoluble in wastewater and leaves the treatment process as escaping bubbles.

Although carbon dioxide is soluble in wastewater, some of the carbon dioxide released by denitrifying bacteria leaves the treatment process as escaping bubbles, if denitrification is severe and carbon dioxide is produced rapidly. Bicarbonate alkalinity is formed when carbon dioxide dissolves in the wastewater.

During intense and sudden episodes of denitrification, two sizes of bubbles can be observed escaping a treatment process. These bubbles are molecular nitrogen and carbon dioxide. Molecular nitrogen is the smaller bubble.

Nitrous oxide, or laughing gas, also is produced and released by denitrifying bacteria. The production and release of nitrous oxide usually occurs under strongly fluctuating conditions or when denitrifying bacteria are adversely affected. The amount of nitrous oxide released is relatively small. Nitrous oxide is insoluble in wastewater and leaves the treatment process as escaping bubbles.

Minute amounts, if any, ammonia and nitric oxide are produced. Both end products are highly toxic to denitrifying bacteria. Ammonia released to the wastewater dissolves to form ammonium ions. Nitric oxide is not ordinarily released from the bacterial cells and does not accumulate in the wastewater. How denitrifying bacteria cope with the presence of nitric oxide is not known.

26

Sources of Nitrite Ions and Nitrate Ions

Unless nitrite ions and nitrate ions are discharged to an activated sludge process in specific industrial waste streams, nitrite ions and nitrate ions that are found in an activated sludge process must be produced in the aeration tank. The production of nitrite ions and nitrate ions is achieved through the biologically mediated reactions of nitrification. The reactions involve the oxidation of ammonium ions and nitrite ions.

Industrial waste streams that contain relatively high concentrations of nitrite ions or nitrate ions are listed in Table 26.1. Nitrification or the production of nitrite ions and nitrate ions in the sewer system does not occur due to unfavorable conditions. These conditions include the presence of large quantities of soluble cBOD, absence of dissolved oxygen or presence of only a low level of dissolved oxygen, short retention time, and relatively small population size of nitrifying bacteria.

Nitrite ions and nitrate ions are very mobile compounds due to their high solubility in water. Because of their mobility, settled solids do not need to be mixed to enhance denitrification.

TABLE 26.1 Industrial Waste Streams Containing Ammonium Ions, Nitrite Ions, and Nitrate Ions

Industrial Waste Steam	Nitrite Ions or Nitrate Ions	Ammonium Ions
Automotive		×
Chemical		×
Coal		×
Corrosion inhibitor	NO_2^-	
Fertilizer		×
Food		×
Leachate		×
Leachate (pretreated)	NO_2^-, NO_3^-	
Livestock		×
Meat		×
Meat (flavoring)	NO_3^-	
Meat (pretreated)	NO_2^-, NO_3^-	
Petrochemical		×
Ordnance		×
Pharmaceutical		×
Primary metal		×
Refinery		×
Steel	NO_2^-, NO_3^-	×
Tannery		×

27

Operational Factors Influencing Denitrification

There are several operational factors that strongly influence denitrification. These factors include the presence of substrate (cBOD), the absence of free molecular oxygen, the presence of an adequate and active population of denitrifying bacteria, pH, temperature, nutrients, and redox potential. The most critical factors are the presence of substrate or readily available carbon and the absence of free molecular oxygen. These two critical factors are reviewed in separate chapters.

Approximately 80% of the bacteria within the activated sludge process are facultative anaerobes and are capable of denitrification. Bacteria in the activated sludge process are present in very large numbers in the bulk solution and MLVSS. Unless the treatment process is experiencing a start-up condition, wash out, toxicity, or recovery from toxicity, an adequate and active population of denitrifying bacteria should be present to ensure denitrification under favorable operational conditions. Denitrifying bacteria can be added to the treatment process by augmenting with commercially prepared bio-augmentation products or seeding with the mixed liquor from another treatment process.

Denitrification can occur over a wide range of pH values. Denitrification is relatively insensitive to acidity but may be slowed at low pH. The range of pH values acceptable for proper floc formation by facultative anaerobes, 6.5 to 8.5, also is acceptable for denitrification. To ensure acceptable enzymatic activity of facultative anaerobe and nitrifying bacteria, the pH in the aeration tank should be maintained

147

at a pH value greater than 7.0. The optimal pH range for denitrification is 7.0 to 7.5.

Because denitrification is biologically mediated, denitrification occurs more rapidly with increasing temperature, and conversely, denitrification occurs more slowly with decreasing temperature. Denitrification is inhibited at wastewater temperature below 5°C. To compensate for decreased denitrification at cold temperature, increasing the MLVSS can increase the number of denitrifying bacteria.

Because denitrification is linked to nitrification, and nitrification also is biologically mediated, warmer temperature favors rapid proliferation of nitrate ions. Warmer wastewater also has less affinity for dissolved oxygen than colder wastewater. Therefore dissolved oxygen is exhausted more easily during warm wastewater conditions, and denitrification occurs more easily during warm wastewater conditions.

Major nutrient needs for facultative anaerobes are nitrogen and phosphorus. Because of the greater energy yield and greater cell production of facultative anaerobes during aerobic respiration of cBOD as compared to anoxic respiration of cBOD, nutrient guidelines for facultative anaerobes during aerobic respiration can be used for these bacteria during anoxic respiration. These guidelines for nitrogen and phosphorus during aerobic respiration are 1.0 mg/l for ammonium ions or 3.0 mg/l for nitrate ions and 0.5 mg/l for orthophosphate ions (HPO_4^{2-}) in the mixed liquor effluent filtrate at all times.

Nitrite ions and nitrate ions are present and used for bacterial degradation of cBOD in an operational condition having a redox potential of +50 to −50 millivolts (mv). Redox is the measurement of the amount of oxidized compounds, such as NO_2^- and NO_3^-, and reduced compounds, such as NH_4^+ in a wastewater sample. Within the range of +50 to −50 millivolts, oxygen is either absence or present in a relatively small quantity, while nitrite ions and nitrate ions are present in relatively large quantities (Table 27.1).

TABLE 27.1 Redox Potential and Respiration

Redox Potential (mv)	Respiration	Electron Acceptor	Condition
>+50 mv	cBOD removal nitrification	O_2	Aerobic or oxic
+50 mv to −50 mv	cBOD removal	NO_2^-, NO_3^-	Anoxic or denitrification

28

Substrate or cBOD

The quantity of substrate or cBOD rather than the quantity of nitrite ions or nitrate ions is considered to be the most important factor that determines denitrification. The larger the quantity of cBOD—especially simplistic, soluble cBOD—the greater is the demand for electron acceptors, such as free molecular oxygen, nitrite ions, and nitrate ions. The greater the demand for electron acceptors is, the greater the chances are that denitrification occurs—even under an aerobic condition.

If the demand for electron acceptors surpasses the supply of free molecular oxygen, the facultative anaerobes or denitrifying bacteria switch their enzymatic machinery from the use of free molecular oxygen as electron acceptors to nitrite ions or nitrate ions as electron acceptors.

Denitrifying bacteria use many ordinary, organic compounds and unusual organic compounds as a source of carbon and energy. Denitrifying bacteria can use organic compounds commonly found in domestic wastewater. Several organic compounds that are added to a denitrification tank to fully denitrify include acetic acid (Equation 28.1), ethanol, glucose, methanol (Equation 28.2), and molasses. Compounds most often added are acetic acid and methanol.

$$NO_3^- + 0.85CH_3COOH + H^+$$

$$\rightarrow 0.1C_5H_7O_2N + 0.45N_2 + 1.2CO_2 + 2.1H_2O \qquad (28.1)$$

(denitrification using acetic acid, CH_3COOH)

$$NO_3^- + 1.8CH_3OH + H^+$$
$$\rightarrow 0.065C_5H_7O_2N + 0.47N_2 + 0.76CO_2 + 2.44H_2O \quad (28.2)$$

$$\text{(denitrification using methanol, } CH_3OH)$$

Methanol is usually the substrate of choice for the addition to a denitrification tank. Methanol is a simplistic, soluble cBOD that quickly enters bacterial cells and is easily degraded.

Complete denitrification or removal of all nitrite ions and nitrate ions usually occurs when a cBOD to nitrite ion and nitrate ion ratio of approximately 3:1 exists. Decreasing this ratio to approximately 3:2 causes breakthrough of nitrite ions and nitrate ions. The operational ratio of 3:1.21 is used for complete denitrification, while a slightly higher ratio of 3:1.05 for settled sewage is used for complete denitrification.

For complete denitrification of nitrate ions, 2.5 mg/l of methanol are required as the substrate per mg/l of nitrate ions present (Table 28.1). For every 2.5 mg/l of methanol degraded or used, 0.5 mg of new cells or sludge are produced (Table 28.1). Within every 0.5 mg of new cells or sludge produced, 0.06 mg of nitrogen can be found as organic nitrogen (Table 28.1). The nitrogen not incorporated into new cells or sludge is released as molecular nitrogen to the atmosphere.

For complete denitrification of nitrite ions, 1.5 mg/l of methanol are required as the substrate per mg/l of nitrite ions present (Table 28.1). For every 1.5 mg/l of methanol degraded or used, 0.3 mg of new cells or sludge are produced (Table 28.1). Within every 0.3 mg of new cells or sludge produced, 0.04 mg of nitrogen can be found as organic nitrogen (Table 28.1). The nitrogen not incorporated into new cells or sludge is released as molecular nitrogen to the atmosphere.

TABLE 28.1 Complete Denitrification of Nitrite Ions and Nitrate Ions Using Methanol

Nitrogenous Ion	Methanol Required per mg/l of Nitrogenous Ion	Cells Produced	Nitrogen in Cells Produced
NO_2^-	1.5 mg/l	0.3 mg	0.04 mg
NO_3^-	2.5 mg/l	0.5 mg	0.06 mg

29

Free Molecular Oxygen

Free molecular oxygen inhibits denitrification by virtue of its competition with nitrite ions and nitrate ions as an electron acceptor for the degradation of cBOD. If free molecular oxygen is in the environment of the bacterial cell and enters the bacterial cell, the cell uses free molecular oxygen (Equation 29.1). The use of free molecular oxygen is preferred over the use of nitrite ions and nitrate ions, because the use of free molecular oxygen yields more cellular energy and cellular growth.

$$C_6H_{12}O_6 + 6O_2 \rightarrow 6CO_2 + 6H_2O \qquad (29.1)$$

The amount of oxygen that inhibits denitrification is relatively small. Concentrations of dissolved oxygen < 1.0 mg/l inhibit denitrification. However, if a dissolved oxygen gradient exists across a floc particle, denitrification does occur in the core of the floc particle; that is, denitrification occurs in the presence of measurable dissolved oxygen (Figure 29.1). Floc particle > 100 μm in size are large enough to produce a dissolved oxygen gradient.

Under a dissolved oxygen gradient, the bacterial cells within the floc particles respire using oxygen, nitrite ions, and nitrate ions at the same time. The bacterial cells at the perimeter of the floc particle use dissolved oxygen for respiration, while the bacteria in the core of the floc particle use nitrite ions and nitrate ions for respiration.

151

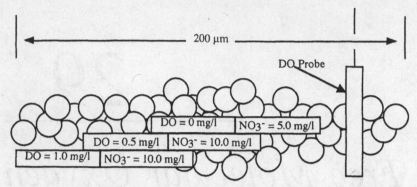

Figure 29.1 *Denitrification in the presence of measurable DO. Although a residual dissolved oxygen level can be measured outside the solids or floc particles in a sludge blanket, denitrification can occur. The occurrence of denitrification in the presence of measurable DO is due to the development of an oxygen gradient from the perimeter to the core of the floc particle. At the perimeter of the floc particle, residual dissolved oxygen is available and is used by the bacteria to degrade cBOD. At the core of the floc particle, residual dissolved oxygen is no longer available, and the bacteria at the core of the floc particle use nitrate ions to degrade cBOD.*

When denitrification occurs, the entire denitrification pathway may not be completed. The denitrification pathway may be shut down at a single oxygen concentration or a significant change in an operational condition. Also some denitrifying bacteria lack key enzymes to complete the entire denitrification pathway.

30

The Occurrence of Denitrification

Denitrification in an activated sludge process can occur whenever appropriate operational conditions exist. These conditions include the presence of an abundant and active population of denitrifying bacteria, an anoxic environment, and the presence of simplistic soluble cBOD. In an anoxic environment a residual amount of free molecular oxygen can be present and measured, but the loss of oxygen across an oxygen gradient results in the use of nitrite ions and nitrate ions at the end of the gradient (Figure 30.1).

Denitrification may be intentional or accidental. Intentional denitrification is the desired use of an anoxic period to achieve a specific operational goal. An anoxic period is established in a denitrification tank in order to satisfy a total nitrogen discharge limit. An anoxic period can be used in an aeration tank to improve floc formation, control undesired growth of filamentous organisms, reduce electrical costs to degrade cBOD, or return the alkalinity to the aeration tank that was lost during nitrification (Figure 30.2).

Accidental denitrification is the undesired occurrence of an anoxic period resulting in increased operational costs, operational problems, and permit violations. Although accidental denitrification most often is observed and reported in secondary clarifiers, accidental denitrification can occur in the sewer system, headworks of the treatment plant, primary clarifiers, chlorine contact tanks, thickeners, and anaerobic digesters (Figure 30.3).

Denitrification problems associated with the sewer system are due

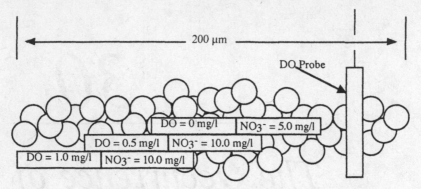

Figure 30.1 *Anoxic environment in the presence of measurable DO. In the presence of an oxygen gradient, an anoxic environment occurs in the core of the floc particle. Operational conditions necessary for the presence of an anoxic environment In the presence of measurable DO include the presence of floc particles that are at least greater than 100 µm in size and the presence of a dissolved oxygen concentration equal to or less than 1.0 mg/l.*

to the discharge of nitrite ions and nitrate ions from specific industries. Denitrification problems associated with the headworks of the treatment plant and the primary clarifiers are due to the discharge of nitrite ions and nitrate ions from specific industries or the recycling of these ions in the RAS to the head of the treatment plant. Denitrification problems associated with the secondary clarifiers, chlorine contact tanks, thickeners, and anaerobic digesters may be due to the discharge of nitrite ions and nitrate ions from specific industries but usually are due to the production of these ions through nitrification in the aeration tanks.

When denitrification occurs in the sewer system, soluble cBOD is rapidly degraded. The degradation of cBOD in the sewer system results in a decrease in the cBOD concentration of the influent wastewater. With a decreased quantity of cBOD in the influent, it becomes more difficult for an activated sludge process to achieve an 85% removal efficiency for cBOD. This difficulty may result in a permit violation for the percentage of cBOD removal.

Denitrification in the sewer system, headworks of the treatment plant, and primary clarifiers reduces the quantity of soluble cBOD or substrate that enters the aeration tank. With reduced substrate in the aeration tank, little bacterial growth occurs. Reduced bacterial growth results in a decrease in MLVSS. Reduced cBOD or substrate may cause many bacteria in the aeration tank to undergo endogenous respiration or die. The occurrence of endogenous respiration

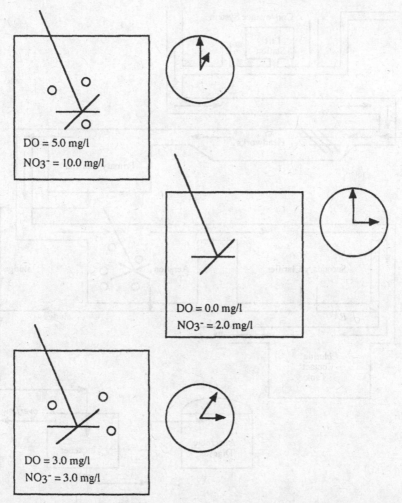

Figure 30.2 *Anoxic period in aeration tank. After the production of nitrate ions (NO_3^-) in an aeration tank, aeration of the tank may be terminated for a period of time, such as one to two hours. During this time period, facultative anaerobic bacteria degrade cBOD using nitrate ions. However, a residual concentration of nitrate ions should be maintained in order to prevent septicity within the aeration tank, and the anoxic period should not be extended beyond four hours. If nitrifying bacteria are deprived of dissolved oxygen for more than four hours, damage to the nitrifying bacterial population may result.*

or death of large numbers of bacteria also results in a decrease in MLVSS.

The death of large numbers of bacteria results in cellular lysis. As bacteria lyze or break open, they release their cellular contents. The contents include ammonium ions. The release of ammonium ions contributes to an elevated level of ammonium ions in the effluent.

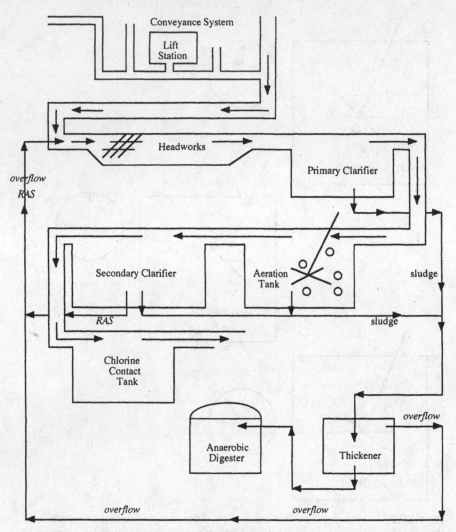

Figure 30.3 *Occurrence of denitrification. Denitrification can occur wherever and whenever an anoxic condition develops. For denitrification to occur, denitrifying bacteria must be present, dissolved oxygen must be absent or a dissolved oxygen gradient must be present, and soluble cBOD must be present. These conditions for the occurrence of denitrification may be found in the conveyance system, headworks of the activated sludge process, primary clarifier, aeration tank during nonaerated periods, secondary clarifier, chlorine contact tank, thickener, and anaerobic digester.*

With a decrease in the quantity of cBOD entering the aeration tanks and reduced MLVSS in the aeration tanks, the activated sludge process becomes highly vulnerable to upsets from slug discharges and toxicity. Also with reduced MLVSS it becomes difficult for an activated sludge process to successfully nitrify.

Denitrification in the primary clarifiers results in the rising of solids to the surface of the clarifiers. These solids must be collected and transferred to appropriate tanks for thickening, digesting, dewatering, and disposal. These operational tasks represent increased costs.

Denitrification in the secondary clarifiers presents several operational concerns. Molecular nitrogen entrapped in the sludge blanket causes a thinning of the sludge blanket and a decrease in the number of bacteria that are returned to the aeration tank in the RAS. Buoyant sludge rising to the surface of the clarifiers represents a loss of solids and bacteria to the receiving water. This loss of solids may represent a permit violation for total suspended solids (TSS).

Coliform bacteria and pathogenic organisms within the solids lost from the secondary clarifiers and discharged to the chlorine contact tanks are protected from disinfection by chlorination. This protection is provided by the presence of solids that surround the organisms and the presence of nitrite ions that may be in the clarifier effluent. The discharge of elevated levels of coliform bacteria to the receiving water may result in a permit violation for coliform bacteria.

Denitrification in the thickener results in poor compaction of solids and floating solids. Poorly compacted solids may require the use of polymers or metal salts to improve compaction. The use of polymers or metal salts represents an increase in operational costs. Poorly compacted solids transferred to an anaerobic digester result in a "washout" of digester alkalinity and a decrease in retention time in the digester. Floating solids due to the entrapment of molecular nitrogen result in an overflow of solids from the thickener to the head of the treatment plant.

Nitrite ions and nitrate ions transferred to an anaerobic digester have two significant and adverse impacts on digester performance. First, the rapid depletion of these ions through denitrification in the digester and the release of molecular nitrogen result in sudden and severe foaming. Second, the presence of nitrite ions and nitrate ions in the digester increases the redox potential of the digester sludge. An increase in redox potential above -300 mv inhibits the activity of methane-forming bacteria that convert volatile acids (cBOD) to methane. Inhibition of methane-forming bacteria permits the accumulation of volatile acids and the production of a "sour" digester.

31

Monitoring and Correcting Accidental Denitrification

The most common occurrence of accidental denitrification is in the secondary clarifier. The occurrence of denitrification in the secondary clarifier is often termed "clumping" or "rising sludge." During denitrification large clumps of dark sludge can be observed rising from the bottom to the top of the clarifier. Numerous bubbles (molecular nitrogen, carbon dioxide, and nitrous oxide) are associated with the sludge. The sludge is dark due to the high MCRT provided for the growth of nitrifying bacteria that produce the nitrite ions and nitrate ions that are used during denitrification.

In the secondary clarifier the bacteria rapidly consume dissolved oxygen as the sludge separates from the clarified supernatant. The loss of dissolved oxygen in the sludge blanket allows for development of an anoxic condition. Under this anoxic condition, nitrite ions and nitrate ions are reduced to molecular nitrogen and nitrous oxide by denitrifying bacteria.

The rate of gas production in the secondary clarifier can vary significantly. Gas production can be extremely high during warm temperatures causing significant turbulence within the clarifier and can inhibit normal settling of solids. Gas production can be very low if an insufficient carbon source (cBOD) is present. Regardless of the rate of gas production, rising sludge is more evident if the floc particles have numerous filamentous organisms. These organisms easily entrap large quantities of gases.

There are several indicators of denitrification. Indicators of denitrification in the secondary clarifier include

- presence of numerous bubbles,
- rising solids that have numerous bubbles on their surface,
- increase in alkalinity, and possibly, pH across the clarifier, and
- reduction in redox potential across the clarifier.

Because the atmosphere contains approximately 80% molecular nitrogen, the capture and measurement of escaping molecular nitrogen from the surface of the clarifier to demonstrate denitrification is not practical. The captured gas may be contaminated by atmospheric molecular nitrogen.

However, capturing and measuring nitrous oxide can be used as an indicator of denitrification. Only a very small quantity of nitrous oxide is present in the atmosphere at low elevations, and the accumulation of nitrous oxide serves as an indicator of denitrification.

Proper mass balance for nitrogen (including the nitrogenous gases released during denitrification) is difficult to make in an activated sludge process. This difficulty is due to the numerous conversions of nitrogen that occur in the activated sludge process. These conversions include not only aerobic nitrification and anoxic denitrification but also assimilatory nitrate reduction, organotrophic nitrification, anaerobic ammonium ion oxidation, and aerobic denitrification (Figure 31.1).

Denitrification usually occurs in the secondary clarifier if the sludge is retained too long, and the sludge blanket becomes deoxygenated. Denitrification in the secondary clarifier also can be associated with clarifier design. Flat-bottom clarifiers with a central sludge takeoff are very susceptible to denitrification. This is due to the accumulation of sludge at the perimeter of the clarifiers.

Denitrification can be controlled in the secondary clarifier by increasing the RAS rate or ensuring that the mixed liquor is well aerated before it enters the clarifier. Design problems contributing to denitrification may be overcome with appropriate baffling.

Denitrification also can be controlled by recycling the RAS to a special anoxic zone at the inlet of the aeration tank where the incoming wastewater itself is used as the carbon source or substrate. The recycled sludge and wastewater are kept in the anoxic zone for a spe-

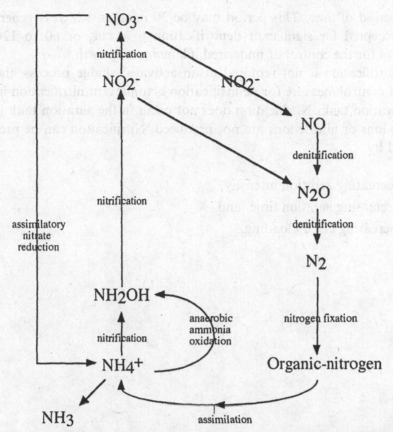

Figure 31.1 *Nitrogen conversions. In the environment or the activated sludge process, nitrogen can be converted from one form of nitrogenous compound to another form of nitrogenous compound. These conversions are due to biological, chemical, and physical events. Nitrate ions (NO_3^-) are perhaps the most significant nitrogenous compounds. Nitrate ions can be used as a nutrient source for nitrogen by many organisms as they undergo assimilatory nitrate reduction to ammonium ions (NH_4^+). Nitrate ions also can be converted to molecular nitrogen (N_2) as they undergo dissimilatory nitrate reduction. Nitrite ions can undergo assimilatory reduction and dissimilatory reduction as they are converted to ammonium ions and molecular nitrogen, respectively. Molecular nitrogen can be fixed or converted to ammonium ions by several nitrogen-fixing bacteria and numerous algae. Once fixed, nitrogen in the form of ammonium ions is easily assimilated into amino acids and proteins—forms of organic nitrogen. Assimilation of ammonium ions in an aeration tank results in an increase in the amount of MLVSS. At relatively high pH values ammonium ions are converted to ammonium (NH_3) and escape from the water to the atmosphere as a gas. Under some anaerobic conditions, ammonium ions can be oxidized to nitrite ions and nitrate ions (anaerobic ammonia oxidation).*

cific period of time. This period may be 30 minutes, which is generally accepted for significant denitrification to occur, or 60 to 120 minutes for the control of undesired, filamentous growth.

If nitrification is not required at an activated sludge process, the easiest control measure for denitrification is to prevent nitrification in the aeration tank. Nitrification does not occur in the aeration tank if nitrite ions or nitrate ions are not produced. Nitrification can be prevented by

· decreasing aeration intensity,
· decreasing aeration time, and
· increasing cBOD loading.

32

Zoning

Desired or intentional denitrification is usually achieved in an activated sludge process by establishing an anoxic zone or an alternating series of anoxic zones (Figure 32.1). The anoxic zone is the first com-

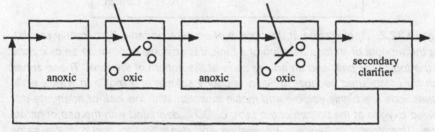

return activated sludge

Figure 32.1 *Alternating series of anoxic zones. In an activated sludge process that has the capability to operate in plug-flow, it is a common practice to control denitrification in the secondary clarifier by using the first aeration tank as an anoxic zone. In this practice, return activated sludge and nitrate ions within the secondary clarifier are removed from the clarifier as rapidly as possible and discharged with the aeration tank influent (primary clarifier effluent) to the first aeration tank. This tank is mixed but not aerated. In the first aeration tank the returned solids (bacteria), returned nitrate ions, and the influent soluble cBOD establish an anoxic condition. Depending on the flow through the first aeration tank, the anoxic zone may vary from one-half to several hours. In the plug-flow mode of operation addition anoxic zones may be placed downstream of aeration tanks that produce nitrate ions. The additional anoxic zones reduce the quantity of nitrate ions discharged to the secondary clarifier.*

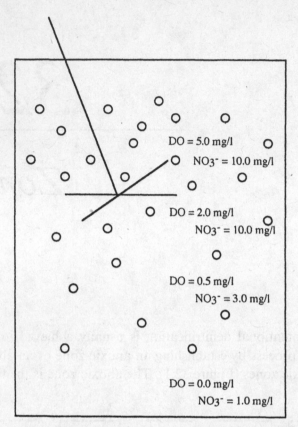

Figure 32.2 *Anoxic zone at the bottom of an aeration tank. By carefully regulating the amount of surface aeration of a tank, it is possible to produce an oxic zone at the top of the tank and an anoxic zone at the bottom of the tank. These zones can be established because oxygen is poorly soluble and motile in water, while nitrate ions are highly soluble and motile in water. With the lack of adequate dissolved oxygen at the bottom of the tank, cBOD is degraded with the use of nitrate ions. Therefore simultaneous nitrification and denitrification occur in the same tank by carefully regulating the amount of surface aeration of a tank.*

partment of the treatment process that denitrifies. Nitrite ions and nitrate ions used for denitrification in the anoxic zone are produced in the aeration tank. The ions enter the anoxic zone through the RAS or aeration tank effluent.

Mixing is provided in the anoxic zone. Slow speed subsurface mixers are used to keep the biomass in suspension and to ensure rapid depletion of any dissolved oxygen.

The size of the anoxic zone provides a retention time of 0.5 to 1 hour. This retention time achieves about 30% to 40% removal of

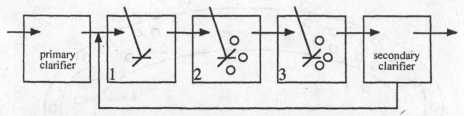

Figure 32.3 *Denitrification in plug-flow mode of operation. In plug-flow mode of operation the first aeration tank is used as a denitrification tank. It is the first tank in plug-flow mode of operation that receives the soluble cBOD required to drive denitrification in the presence of nitrate ions and denitrifying bacteria. The nitrate ions and denitrifying bacteria enter the first aeration tank through the solids returned to the tank from the secondary clarifier. The first aeration tank is not aerated when it is used as a denitrification tank.*

nitrogen. Because the rate of denitrification declines with increasing retention time, larger tanks providing greater retention times are not economical to operate. However, anoxic retention times of 1 to 2 hours are needed to control the undesired growth of filamentous organisms.

Although the anoxic zone is most effective if it is in a separate tank, denitrification can occur in a tank that is required to perform an additional biological role, such as nitrification. Denitrification stops in the anoxic zone if the denitrifying bacteria are exposed to dissolved oxygen. Nitrification stops if the dissolved oxygen concentration in contact with the nitrifying bacteria is too low.

There are several means to achieve zoning. Surface aeration of a mixed liquor tank can be adjusted to provide an aerobic zone at the surface of the tank and an anoxic zone at the bottom of the tank (Figure 32.2). Denitrification also can be achieved in a plug-flow mode of operation (Figure 32.3). In this mode of operation the anoxic zone is placed in the front of the system. Here the zone has the largest amounts of nitrite ions and nitrate ions and the largest amount of soluble cBOD.

Denitrification can be achieved in oxidation ditches (Figure 32.4). Oxidation ditches have a high degree of natural internal recirculation that makes denitrification affordable and inexpensive compared to the plug-flow mode of operation. An oxidation ditch can be regulated to establish discrete aerobic zones and anoxic zones along the channels.

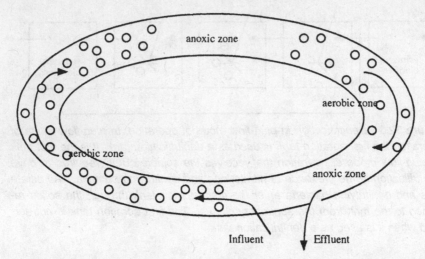

Figure 32.4 *Denitrification in an oxidation ditch. The design and internal mixing of an oxidation ditch permit the establishment of aerated and non-aerated zones with the oxidation ditch for nitrification and denitrification. Aerated zones that produce nitrate ions are followed by non-aerated zones that use the nitrate ions for degradation of soluble cBOD.*

Zoning also can occur in an aeration tank when the tank is poorly mixed. Under this condition anoxic pockets are produced where denitrification can occur. Zoning can be mimicked in sequential batch reactors (SBR) by extending the aeration time to achieve nitrification and extending the settling time to achieve denitrification.

33

Benefits of Denitrification

The benefits of denitrification are significant if denitrification is used properly. The benefits include protecting the quality of the receiving water, permit compliance, strengthening of the floc particles, control of undesired filamentous growth, return of alkalinity to the treatment process, and cost-savings for the treatment or degradation of cBOD.

By denitrifying, the quantity of nitrite ions and nitrate ions discharged to the receiving water is greatly reduced. The ions are used as a nitrogen nutrient by many aquatic plants. The reduction in the quantity of these ions discharged to the receiving water helps to prevent the overabundant growth of aquatic plants and its resulting eutrophication.

Nitrite ions are highly toxic to aquatic life. Denitrification reduces the amount of nitrite ions discharged to the receiving water. The reduction of nitrite ions discharged to the receiving water reduces toxicity concerns related to aquatic life.

Activated sludge processes that have a total nitrogen discharge limit must denitrify. Successful denitrification ensures permit compliance.

Denitrification promotes the formation of firm and dense floc particles. Firm and dense floc particles are resistant to shearing action and have desired settling characteristics. An anoxic environment produced through denitrification favors the growth of facultative anaerobic, floc-forming bacteria and discourages the growth of strict aerobic, filamentous organisms and weak facultative anaerobic, filamentous organisms. Strict aerobic, filamentous organisms can only use free molecular oxygen. Therefore their growth is stopped, or the organisms die under an anoxic environment.

Although weak facultative, anaerobic filamentous organisms can use free molecular oxygen and nitrite ions or nitrate ions, these filamentous organisms cannot compete successfully for nitrite ions or nitrate ions with facultative anaerobic, floc-forming bacteria. Therefore the growth of these filamentous organisms is stopped, or the organisms die under an anoxic environment.

Nearly all activated sludge processes that denitrify must nitrify. Nitrification results in a loss of alkalinity. Denitrification returns alkalinity to the activated sludge process. By denitrifying, much of the alkalinity lost during nitrification can be returned to the activated sludge process through denitrification.

By using nitrite ions and nitrate ions produced during nitrification to degrade cBOD, an activated sludge process need not supply dissolved oxygen to the aeration tank. The use of nitrite ions and nitrate ions instead of dissolved oxygen results in cost-savings, that is, decreased electrical cost for operation of the aeration equipment.

The Gram Stain

The Gram stain uses a simplistic staining technique to differentiate bacteria into two large groups, Gram-negative bacteria and Gram-positive bacteria. The basis for this differentiation is due to the physical and chemical composition of the cell walls of bacteria to retain and release stains.

Christain Gram, a Danish physician, who was working on a method to differentiate pathogenic bacteria from mammalian tissue, developed the Gram stain in 1883. All bacteria, except one genus, respond to the Gram stain technique. Following staining, Gram-negative bacteria are pink-red when examined under the microscope, while Gram-positive bacteria are blue-purple when examined under the microscope.

There are several Gram staining techniques. The technique most commonly used in wastewater laboratories for the identification of different bacteria is the Hücker method. This method employs the use of four reagents. These reagents consist of crystal violet, Gram's iodine, a decolorizing agent, and safranin. These reagents can be purchased from most chemical suppliers to water and wastewater laboratories. The stains are applied to a smear of bacteria on a microscope slide in the following order: crystal violet, Gram's iodine, decolorizing agent, and safranin. The color of Gram-negative and Gram-positive bacteria after the application of each reagent is presented in Table I.1. The Gram stain results of the bacteria are observed after the safranin addition.

TABLE I.1 Bacterial Response to Each Reagent of the Gram Stain Technique

Reagent	Gram-Negative Bacteria	Gram-Positive Bacteria
Crystal violet	Blue-purple	Blue-purple
Gram's iodine	Blue-purple	Blue-purple
Decolorizing agent	Colorless	Blue-purple
Safranin	Pink-red	Blue-purple

Appendix II

F/M, HRT, MCRT

FOOD/MICROORGANISM RATIO (F/M) CALCULATION

The food/microorganism ratio, or F/M, is a measurement of the food entering the activated sludge process and the microorganisms in the aeration tank(s). Each activated sludge process has an F/M at which it operates best. This F/M may fluctuate throughout the year according to changes in operational conditions, such as industrial discharges, permit requirements, and temperature.

The food value or food supply entering the activated sludge process consists of the BOD loading or pounds discharged to the aeration tank(s). The BOD loading is calculated by multiplying the concentration (mg/l) of BOD entering the aeration tank by the influent aeration tank flow in millions of gallons per day (MGD) by the weight constant of 8.34 pounds per gallon of wastewater (Equation II.1).

BOD mg/l × Flow (MGD) × 8.34 pounds/gal wastewater

= BOD loading (II.1)

The microorganism value or amount of microorganisms in the activated sludge process consists of the pounds of mixed liquor volatile suspended solids (MLVSS) in the on-line aeration tank(s). The pounds of MLVSS is calculated by multiplying the concentration (mg/l) of MLVSS by the aeration tank(s) volume in million gallons

(MG) by the weight constant of 8.34 pounds per gallon of waste-water (Equation II.2).

$$\text{MLVSS (mg/l)} \times \text{Aeration tank volume (MG)}$$

$$\times \text{ 8.34 pounds/gal wastewater} = \text{pounds MLVSS} \qquad \text{(II.2)}$$

The F/M of an activated sludge process can be calculated by dividing the pounds of food as BOD applied to the microorganisms or MLVSS present in on-line aeration tanks (Equation II.3)

$$\text{F/M} = \text{Pounds BOD to aeration tank/Pounds MLVSS in}$$
$$\text{aeration tank} \qquad \text{(II.3)}$$

HYDRAULIC RETENTION TIME (HRT) CALCULATION

The hydraulic retention time or HRT is the amount of time in hours for wastewater to pass through a tank, such as an aeration tank. Changes in the HRT of an activated sludge process can affect biological activity. For example, decreasing HRT adversely affects nitrification, while increasing HRT favors nitrification and the solublization of colloidal BOD and particulate BOD.

The HRT of an aeration tank is determined by dividing the volume of the aeration tank in million gallons by the flow rate through the aeration tank (Equation II.4). The flow rate through the aeration tank must be expressed as gallons per hour (gph).

$$\text{HRT (hours)} = (\text{Volume of aeration tank, gal})/(\text{Flow rate, gph})$$
$$\text{(II.4)}$$

MEAN CELL RESIDENCE TIME (MCRT) CALCULATION

The mean cell residence time or MCRT is the amount of time, in days, that solids or bacteria are maintained in the activated sludge system. The MCRT is known also as the solids retention time (SRT). To calculate the MCRT of an activated sludge process, it is necessary to know the amount or pounds of solids or suspended solids in the activated sludge system and the amount or pounds of suspended solids leaving the activated sludge system.

To determine the pounds of suspended solids in the activated sludge system, the pounds of mixed liquor suspended solids (MLSS) must be calculated. The MLSS consists of all solids in the aeration tank(s) and secondary clarifier(s). Therefore the pounds of MLSS in an activated sludge systems consists of the concentration (mg/l) of MLSS times the volume (MG) of the aeration tank(s) and clarifier(s) times the weight constant of 8.34 pounds per gallon of wastewater (Equation II.5).

Pounds of MLSS

$$= \text{MLSS mg/l} \times (\text{Volume of aeration tanks} + \text{Clarifiers, MG})$$

$$\times 8.34 \text{ pounds/gal wastewater} \qquad (\text{II.5})$$

To determine the pounds of suspended solids leaving the activated sludge process, the amount or pounds of suspended solids loss through wasting and discharge in the secondary effluent must be calculated. Therefore the pounds of suspended solids leaving the activated sludge process consists of pounds of activated sludge wasted per day and the pounds of activated sludge or secondary effluent suspended solids discharged per day (Equation II.6).

Pounds of suspended solids leaving activated sludge process

$$= \text{Wasted sludge (mg/l)} \times \text{Wasted sludge flow (MGD)}$$

$$\times 8.34 \text{ pounds/gal wastewater}$$

$$+ \text{Secondary effluent suspended solids (mg/l)}$$

$$\times \text{Effluent flow (MGD)} \times 8.34 \text{ pounds/gal wastewater} \quad (\text{II.6})$$

The mean cell residence time of an activated sludge process can be calculated by dividing the pounds of suspended solids or MLSS in the activated sludge system by the pounds per day of suspended solids leaving the activated sludge system (Equation II.7).

$$\text{MCRT} = \frac{\text{Suspended solids in system, pounds}}{\text{Suspended solids leaving system per day}} \qquad (\text{II.7})$$

References

Abufayed, A. A., and E. D. Schroeder. 1986. Kinetics and stoichiometry of SBA/denitrification with a primary sludge carbon source. *J. Wat. Control Fed.* (5).

Austin, B., ed. 1988. *Methods in Aquatic Bacteriology.* Wiley, New York.

Bitton, G. 1994. *Wastewater Microbiology.* Wiley-Liss, New York.

Canter, L. W. 1997. Nitrates in Groundwater. CRC, Lewis Publishers, Boca Raton, FL.

Chudoba, J., J. S. Cech, and P. Chudoba. 1985. The effect of aeration tank configuration on nitrification kinetics. *J. Wat. Control Fed.* (11).

Celenza, Gaetano J. 2000. *Industrial Waste Treatment Process Engineering; Biological Processes*, Vol. 2. Technomic Publishing, Lancaster.

Delwiche, C. C., ed. 1981. *Denitrification, Nitrification, and Atmospheric Nitrous Oxide.* Wiley, New York.

Dinges, R. 1982. *Natural Systems for Water Pollution Control.* Van Nostrand Reinhold, New York.

Eckenfelder, W. W. 1997. *Developing Industrial Water Pollution Control Programs, A Primer.* Technomic Publishing, Lancaster.

Eckenfelder, W. W., and P. Grau, eds. 1992. *Activated Sludge Process Design and Control.* Technomics Publishing, Lancaster.

Ferrris, T. C., chairman. 1983. *Manual of Wastewater Treatment.* The Texas Water Utilities Association, Austin.

Galli, E., S. Silver, and B. Witholt. 1992. *Pseudomonas, Molecular Biology and Biotechnology.* Am. Soc. Microbiol, Washington, DC.

Ganczarczyk, J. J. 1983. *Activated Sludge Process, Theory and Practice.* Dekker, New York.

Gerardi, M. H., chairman. 1991. *Wastewater Biology: The Microlife*. Water Environment Federation. Alexandria, VA.

Gerardi, M. H., chairman. 1994. *Wastewater Biology: The Life Processes*. Water Environment Federation. Alexandria, VA.

Gerardi, M. H., chairman. 2001. *Wastewater Biology: The Habitats*. Water Environment Federation. Alexandria, VA.

Gottschalk, G. 1986. *Bacterial Metabolism*. Springer, New York.

Gray, N. F. 1990. *Activated Sludge; Theory and Practice*. Oxford University Press. Oxford.

Hall, E. R., and K. L. Murphy. 1985. Sludge age and substrate effects on nitrification kinetics. *J. Wat. Control Fed.* (5).

Haller, E. J. 1995. *Simplied Wastewater Treatment Plant Operations*. Technomics Publishing, Lancaster.

Hart, J. E., R. Defore, and S. C. Chiesa. 1986. Activated sludge control for seasonal nitrification. *J. Wat. Control Fed.* (5).

Huang, J. M., J. H. Oliver, Y. C. Wu, and A. H. Molof. 1989. Nitrification of activated sludge effluent in a cross-flow medium trickling filter. *J. Wat. Control Fed.* (4).

Irvine, R. L., G. Miller, and A. S. Bhamrah. 1979. Sequencing batch treatment of wastewaters in rural areas. *J. Wat. Control Fed.* (2).

Iwai, S., and T. Kitao. 1999. Wastewater treatment with microbial films. Technomic Publishing, Lancaster.

Johnston, D. O., J. T. Netterville, J. L. Wood, and M. M. Jones. 1973. *Chemistry and the Environment*. Saunders, Philadelphia, PA.

Kadlec, R. H., and R. L. Knight. 1996. *Treatment Wetlands*. CRC, Lewis Publishers. Boca Raton, FL.

Linne, S., S. Chiesa, M. Rieth, and R. Polta. 1989. The impact of selector operation on activated sludge settleability and nitrification: Pilot-scale results. *J. Wat. Control Fed.* (1).

Manahan, S. E. 1994. *Environmental Chemistry*, 6th ed. Lewis Publishers. Boca Raton, FL.

Menoud, P., C. H. Wong, H. A. Robinson, A. Farquhar, J. P. Barford, and G. W. Barton. 1999. Simultaneous nitrification and denitrification using Siporax^{TM} packing. *Wat. Sci. Tech.* (40).

Mitchell, R. ed. 1972. *Water Pollution Microbiology*. Wiley-Interscience, New York.

Mudrack, K., and S. Kunst. 1986. *Biology of Sewage Treatment and Water Pollution Control*. Wiley, New York.

Newton, J. J. 1985. Special report: Chemicals for wastewater treatment. *Pollution Eng.* (11).

Newton, W. E., and W. H. Orme-Johnson. 1978. *Nitrogen Fixation*; Vol. 1: *Free-living Systems and Chemical Models*. University Park Press, Baltimore.

O'Grady, T. 1994. Meeting new nitrification requirements at existing activated sludge treatment plants. *Keystone Wat. Qual. Manager.* (2).

Palis, J. C., and R. L. Irvine. 1985. Nitrogen removal in a low-load single tank sequencing batch reactor. *J. Wat. Control Fed.* (1).

Pochana K., and J. Keller. 1999. Study of factors affecting simultaneous nitrification and denitrification (SND). *Wat. Sci. Tech.* (39).

Rowe, D. R., and I. M. Abdel-Magid. *Handbook of Wastewater Reclamation and Reuse*. CRC Lewis Publishers, Baca Raton, FL.

Sedlak, R., ed. 1991. *Phosphorus and Nitrogen Removal from Municipal Wastewater-Principles and Practices*, 2nd. Lewis Publishers, Boca Raton, FL.

Shammas, N. 1986. Interactions of temperature, pH, and biomass on the nitrification process. *J. Wat. Control Fed.* (1).

Spector, M. 1998. Cocurrent biological nitrification and denitrification in wastewater treatment. *Wat. Env. Res.* (70).

Suzuki, I., S. Kwok, and U. Dular. 1976. Competitive inhibition of ammonia oxidation in nitrosomonas europaea by methan, carbon monoxide or methanol. *FEFBS Letters* (72).

Thiem, L. T., and E. A. Alkhatib. 1989. In situ adaptation of activated sludge by shock loading to enhance treatment of high ammonia content petrochemical wastewater. *J. Wat. Control Fed.* (7).

Turk, O., and D. S. Mavinic. 1989. Stability of nitrite build-up in an activated sludge system. *J. Wat. Control Fed.* (8).

Van De Graaf, A. A., A. Mulder, P. DeBruun, M. S. M. Jetten, L. A. Robertson, and J. G. Kuenen. 1995. Anaerobic oxidation of ammonium in a biological mediated process. *App. Env. Micro.* (61).

Van Loosdrecht, M. C. M., and M. S. M. Jetten. 1998. Microbiological conversions in nitrogen removal. *Wat. Sci. Tech.* (38).

Wilson, R. W., K. L. Murphy, P. M. Sutton, and S. L. Lackey. 1981. Design and cost comparison of biological nitrogen removal processes. *J. Wat. Control Fed.* (8).

Abbreviations and Acronyms

ATP	Adenosine triphosphate
BOD	Biochemical oxygen demand
cBOD	Carbonaceous biochemical oxygen demand
coBOD	Colloidal biochemical oxygen demand
DO	Dissolved oxygen
F/M	Food to microorganism ratio
HRT	Hydraulic retention time
I/I	Inflow and infiltration
kcal	Kilocalories
MCRT	Mean cell residence time
mg	Milligram
mg/l	Milligrams per liter
MLSS	Mixed liquor suspended solids
MLVSS	Mixed liquor volatile suspended solids
mv	Millivolt
nBOD	Nitrogenous biochemical oxygen demand
nm	Nanometer
NOD	Nitrogenous oxygen demand
pBOD	Particulate biochemical oxygen demand
RAS	Return activated sludge
sBOD	Soluble biochemical oxygen demand

SBR	Sequential batch reactor
tBOD	Total biochemical oxygen demand
TKN	Total kjeldahl nitrogen
TSS	Total suspended solids
μm	Micron
°C	Degrees Celsius
#	Pound or number
>	Greater than
<	Less than

Chemical Compounds and Elements

C	Carbon
$CaCO_3$	Calcium carbonate
$Ca(HCO_3)_2$	Calcium bicarbonate
$Ca(OH)_2$	Calcium hydroxide
$CH_3CH_2CH_2OH$	n-propanol
$(CH_3)_2CHOH$	i-propanol
$CH_3(CH_2)_{16}COOH$	Stearic acid
CH_3CH_2OH	Ethanol
$CH_3CHOHCH_3$	Isopropyl alcohol
CH_3COCH_3	Acetone
$CH_3CO_2C_2H_5$	Ethyl acetate
$(CH_3)_3COH$	t-propanol
$C_6H_{12}O_6$	Glucose
CH_3COOH	Acetic acid
CH_2NH_2	Methylamine
$CH_3NH_2CH_2OH$	Aminoethanol
CH_3OH	Methanol
$C_5H_7O_2N$	Cellular material
Cl^-	Chloride ion
CO_2	Carbon dioxide
CO_3^{2-}	Carbonate
$-COOH$	Carboxylic acid (carboxyl) group

$CuSO_4$	Copper sulfate
H	Hydrogen
H^+	Hydrogen ion
HCO_3^-	Bicarbonate ion
H_2CO_3	Carbonic acid
$HgCl_2$	Mercuric chloride
HNO_2	Nitrous acid
HOCl	Hypochlorous acid
$MgCO_3$	Magnesium carbonate
$Mg(HCO_3)_2$	Magnesium bicarbonate
$Mg(OH)_2$	Magnesium hydroxide
N	Nitrogen
N_2	Molecular nitrogen
$-NH_2$	Amino group
NH_3	Ammonia
NH_4^+	Ammonium ion
NH_2CONH_2	Urea
NH_4HCO_3	Ammonium bicarbonate
NH_2OH	Hydroxylamine
NH_4OH	Aqua ammonia
NO	Nitric oxide
N_2O	Nitrous oxide
NO_2^-	Nitrite ion
NO_3^-	Nitrate ion
NOH	Nitroxyl
Na_2CO_3	Sodium carbonate
$NaHCO_3$	Sodium bicarbonate
$NaHSO_3$	Sodium bisulfite
NaOH	Sodium hydroxide
O_2	Free molecular (dissolved) oxygen
OCl^-	Hypochlorite ion
OH^-	Hydroxyl ion
PO_4^{2-}	Phosphate
$-SH$	Thiol group
SO_2	Sulfur dioxide
SO_4^{2-}	Sulfate

Glossary

absorb Penetration of a substance into the body of an organism

acclimate Gradual repair or replacement of enzymes damaged by inhibitory compounds

actinomycete Filamentous bacteria, moldlike bacteria

acute Having a sudden onset and short course

adsorb The taking up of one substance at the surface of an organism

aerobic The use of free molecular oxygen for cellular respiration

aggregate Crowded or massed into a dense cluster

alkalinity Having a pH greater than 7

alkalis Chemical compound that releases alkalinity in water

allyl alcohol 1-hydroxy prop-2-ene, $H_2C=CHCH_2OH$, an unsaturated primary alcohol, present in wood spirit, made from glycerin and oxalic acid

amino acid A group of organic acids in which a hydrogen atom of the hydrocarbon (alkyl) radical is exchanged for the amino group; used in the production of proteins

ammonification The release of amino groups or ammonia from organic-nitrogen compounds by microbial activity

anaerobic An environment where free molecular oxygen is not used by bacteria for the degradation of substrate

analine Phenylamine, $C_6H_5NH_2$, a colorless oily liquid that is slightly soluble in water; basis for the manufacture of dyestuffs, pharmaceutical, plastic, and many other products

Ångstrom Named for Swedish physicist, A. J. Angstrom (1814–1874); unit of wavelength for electromagnetic radiation covering visible light and X rays; equal to 10^{-10} m.

anoxic An environment where bacteria use nitrite ions or nitrate ions

aqueous Relating to or made with water

asexual Without sex, lacking, or apparently lacking functional sexual organs

assimilatory A general term for all the metabolic processes that permit the buildup of nutrients utilized by organisms

atom The smallest particle of an element that can take part in a chemical reaction

bacillus A rod-shaped bacterium or a genus in the family Bacillaceae

bactericide A substance capable of killing bacteria

binary fission A process in which two similarly sized and shaped cells are formed by the division of one cell; process by which most bacteria reproduce

bioaugmentation The addition of commercially prepared cultures of organotrophs and nitrifying bacteria to a wastewater treatment process to improve operational conditions

biochemical A chemical reaction occurring inside a living cell

biological holdfast A series of lacelike threads providing fixed film, bacterial growth that is immersed in a suspended growth system or activated sludge process

biomass The quantity or weight of all organisms within the treatment process

biosynthetic pathway A series of biochemical reactions resulting in the production of complex molecules in an organism

brackish water Water having less salt than seawater, but undrinkable

bristleworm An aerobic, multicellular, segmented worm with very stiff, erect hairs

budding A form of asexual reproduction in which a daughter cell develops from a small outgrowth or protrusion of the parent cell; the daughter cell is smaller than the parent cell

cationic A compound or material having a net positive charge

cellulose   A polysaccharide consisting of numerous glucose molecules linked together to form an insoluble starch

chelating agent   An organic compound in which atoms form more than one coordinate bond with metals in solution

chloramine   A compound containing chlorine that is substituted for hydrogen; used in the disinfection of potable water supplies

chlorinated hydrocarbon   An organic compound having chlorine substituted for at least one hydrogen atom

chronic   Long term or duration

ciliated protozoa   Single-celled organism having short, hairlike structures or cilia

coccus   A spherical-shaped bacterium

coliform   Gram-negative, lactose-fermenting, enteric rod-shaped bacteria, such as *Escherichia coli*

colloid   Suspended solid with a large surface area that cannot be removed by sedimentation alone

colorimetric   The use of an instrument for the precise measurement of the hue, purity, and brightness of a color

cytomembrane   Internal membrane found immediately beneath the cell membrane of nitrifying bacteria; the active site for the oxidation of ammonium ions and nitrite ions

deaminase   Enzyme specific for the removal of amino groups through bacterial activity

deaminate   The release of amino groups through microbial activity

degrade   The use of enzymes to breakdown or oxidize BOD resulting in the release of energy to bacterial cells

diffusion   The spatial equalization of one material throughout another

disinfectant   An agent that kills or causes a reduction in number of pathogens in wastewater

dissociate   The reversible or temporary breaking-down of a molecule into simpler molecules or atoms

dissimilatory   The reduction of nitrite ions or nitrate ions to molecular nitrogen for the degradation of cBOD

electron   A fundamental particle with negative charge; electrons are grouped round the nuclei of atoms in several possible orbits

endogenous   The degradation of internal reserve substrate

enumerate   To count

enzyme A proteinaceous molecule found inside a cell or released by a cell that expedites the rate of a biochemical reaction without being consumed in the reaction

exoenzyme An enzyme produced within the cell and released to the environment; numerous exoenzymes solublize pBOD and coBOD that come in contact with the cell

fermentation A mode of energy-yielding metabolism that involves a sequence of oxidation-reduction reactions of two organic compounds

filtrate The liquid and its contents that pass through filter paper

flatworm An aerobic, multicellular "flat" worm having a flat ventral surface and a curved dorsal surface

food chain Scheme of feeding relationships by trophic levels, which unites the member species of a biological community

food web An interrelationship among organisms in which energy is transferred from one organism to another; each organism consumes the preceding one and in turn is eaten by the following member in the sequence

free-living Living or moving independently

fungus A member of a diverse group of unicellular and multicellular organisms, lacking chlorophyll and usually bearing spores and often filaments

generation time The time required for the cell population or biomass to double

genetic material Nucleic materials passed from the parent cell to the daughter cell that contains the code or "instructions" related to the future characteristics and development of the daughter cell

genus A taxonomic or classification group of organisms above the species level that share many similar features

habitat The part of the physical environment in which an organism lives

heavy metal A metal that can be precipitated by hydrogen sulfide in an acid solution and that may be toxic to the activated sludge biomass

hydantoin Diketotetrahydroglyoxaline, $CH_2NH(CO)_2NH$, a naturally occurring carbon- and nitrogen-containing ring compound produced from the degradation of proteins

hydrolysis The biochemical process of decomposition involving the splitting of a chemical bond and the addition of water

infiltration   Groundwater that enters the sewer system through cracks in laterals, mains, and manholes

inflow   Strom water that enters the sewer system through catch basins and downspouts

inhibition   The act of repressing enzymatic activity

inorganic   Compounds that do not contain the elements carbon or hydrogen

intermediate   The precursor to a desired product

lysis   To break open; namely upon the death of bacterial cells, the content of the cells is released to the environment

mercaptan   Thio-alcohols; contain –SH; they form salts with sodium, potassium, and mercury, and are formed by warming alkyl halides or sulfates with potassium hydrosulfide in concentrated alcoholic or aqueous solution

metabolism   Pertaining to cellular activity, such as the degradation of BOD

metal salt   A coagulant, such as alum, ferric chloride, or lime, used for solids capture, dewatering, or thickening

micron   Measure of length equal to one millionth of a meter or one thousandth of a milliliter

molecule   Smallest part of a compound that exhibits all the chemical properties of that specific compound

multicellular   Many cells

nanometer   A measurement of length; one thousandth of a micron

nematode   Any member of a group of nonsegmented worms

nitrogenase   Enzyme specific for the breaking of the triple bond in molecular nitrogen

nodule   A small rounded structure on the root of a plant inhabited by symbiotic bacteria

nucleic acid   A large, acidic, chainlike macromolecule containing phosphoric acid, sugar, and purine and pyrimidine bases

obligate   Required

oxic   An environment where bacteria use free molecular oxygen

oxidation   The biological or chemical addition of oxygen to a compound

oxidation state   The number of electrons that must be added to a cation or removed from an anion to produce a normal atom

particulate Insoluble material that may or may not be biodegradable

pathogen A disease-causing agent

pepidogylcan The rigid component of the cell wall in most bacteria, consisting of a gylcan (sugar) backbone

peripheral Situated or produced around the edge

phenol Carbolic acid, C_6H_5OH, chief constituent of coal-tar

photosynthetic The process in which radiant (light) energy is adsorbed by specialized pigments of a cell and is subsequently converted to chemical energy

physiological Pertaining to the functions of living organisms and their physiochemical parts and metabolic reactions

polyphosphate Inorganic compound in which two or more phosphorus atoms are joined together by oxygen

potable water Drinking water supply

protein A class of high molecular weight polymers composed of amino acids joined by peptide linkages

reduction The addition of electrons from a compound; the removal of oxygen

respiration A mode of energy-yielding metabolism that requires a final electron carrier for substrate oxidation

redox The measurement of the amount of oxidized compounds and reduced compounds in an environment

respiration A mode of energy-yielding metabolism that requires a final electron carrier for substrate oxidation

rotifer An aerobic, multicellular, nonsegmented animal having two bands of cilia surrounding a mouth opening; the cilia provide locomotion and a mechanism for gathering food, such as dispersed bacteria

selective agar A dried mucilaginous substance with marked gel-forming properties obtained from oriental seaweed, used in the preparation of various media for the culture of specific bacteria

skatol 3-methylindole; possess a foul odor

slug discharge Usually two and one-half or three times the normal or expected loading, such as BOD

soluble In solution

substrate Food or BOD

supernatant The liquid above settled solids

symbiotic An obligatory interactive association between members of two populations, producing a stable condition in which the two organisms live together in close physical proximity, to their mutual benefit

thiourea Thiocarbamide, NH_2CSNH_2; it is slightly soluble in water, ethanol, and ethoxyethane; used in organic synthesis and as a reagent for bismuth

total kjeldahl nitrogen The amount of ammonia and organic nitrogen within a wastewater sample

unicellular One cell

urease Enzyme specific for the degradation or urea

ultraviolet radiation Short wavelengths of electromagnetic radiation in the range of 100 to 400 nm

valence Oxidation state or charge

waterbear An aerobic, multicellular, invertebrate having four pair of legs with claws

xenobiotic A synthetic product that is not formed by natural biosynthetic processes; a foreign substance or poison

Index

Acclimate, 119
Actinomycetes, 39
Air stripping of ammonia, 67, 73, 80, 88
Algae, 4, 5, 6, 39, 64, 65, 66, 87, 88, 161
Alkalinity, 54, 56, 88, 109–114, 121,
    129, 130, 141, 143, 157, 160, 167,
    168
Amino acid, 18–19, 20, 64, 65, 66, 67,
    68, 78, 79, 88, 101, 161
Ammonia discharge limit, 3
Ammonia toxicity, 4
Ammonification, 67
Anaerobes, 103, 104
Analine, 17
Anoxic condition, 69, 70, 71, 106, 135,
    138, 139, 153, 154, 155, 156, 159,
    160, 163, 164, 165, 167
Assimilatory nitrate reduction, 79, 136,
    160, 161
ATP, 27, 28

Bacterial
    Capsule, 22
    Cell membrane, 21, 24, 78
    Cell wall, 21, 24, 78, 169
    Cytoplasm, 20, 21, 22, 24
    Flagella, 20, 22, 25
    Mesosomes, 21, 24
    Mitochondria, 22
    Ribosomes, 21, 22, 24
    Storage granules, 21, 24
Bicarbonates, 110, 111, 112, 113
Binary fission, 48, 49, 51
Bioaugmentation products, 130, 131,
    147
Biological holdfast system, 130, 131
Bristleworms, 97, 99, 100
Budding, 48, 49, 51

Carbonates, 110, 111, 112, 113
Cellulose, 60, 96, 101
Chelating agents, 17
Chloramine, 71
Chlorination, 84, 91, 157
Chlorine demand, 88, 91
Chlorine sponge, 91, 92, 93–94, 138
Clumping, 88, 159
Coliform bacteria, 8, 45, 88, 92, 94, 157
Colloids, 19, 21, 25, 55, 58, 73, 95, 96,
    101, 172
Corrosion inhibitors, 17
Cyanide, 120
Cytomembranes, 48, 49

Dark solids, 89
Deamination, 18, 19, 20, 33, 66, 67, 68,
    74, 84, 88, 112, 121
Deaminating bacteria, 19